RIVER WATER QUALITY MONITORING

By Larry W. Canter

74438

LEWIS PUBLISHERS, INC.

Library of Congress Cataloging in Publication Data

Canter, Larry W.
    River water quality monitoring

    "July, 1984"
    Bibliography: p.
    Includes index.
    1. Water quality management. 2. Rivers. I. Title
TD365.C38   1985   363.7'3942'072   84-29702
ISBN 0-87371-011-8

LEWIS PUBLISHERS, INC.
121 South Main Street, Chelsea, Michigan 48118

PRINTED IN THE UNITED STATES OF AMERICA

Larry W. Canter

LARRY W. CANTER, P.E., is the Sun Company Professor of Ground Water Hydrology, and Director, Environmental and Ground Water Institute, at the University of Oklahoma, Norman, Oklahoma, in the USA. Dr. Canter received his Ph.D. in Environmental Health Engineering from the University of Texas in 1967, MS in Sanitary Engineering from the University of Illinois in 1962, and BE in Civil Engineering from Vanderbilt University in 1961. Before joining the faculty of the University of Oklahoma in 1969, he was on the faculty at Tulane University and was a sanitary engineer in the U.S. Public Health Service. He served as Director of the School of Civil Engineering and Environmental Science at the University of Oklahoma from 1971 to 1979.

Dr. Canter has published several books and has written chapters in other books; he is also the author or co-author of numerous papers and research reports. His research interests include environmental impact assessment and ground water pollution control. In 1982 he received the Outstanding Faculty Achievement in Research Award from the College of Engineering, and in 1983 the Regent's Award for Superior Accomplishment in Research.

Dr. Canter currently serves on the U.S. Army Corps of Engineers Environmental Advisory Board. He has conducted research, presented short courses, or served as advisor to institutions in Mexico, Panama, Colombia, Venezuela, Peru, Scotland, The Netherlands, France, Germany, Italy, Greece, Turkey, Kuwait, Thailand, and the People's Republic of China.

PREFACE

River water quality monitoring is often required for establishing baseline conditions, setting quality criteria and standards, monitoring temporal changes, and determining the impacts of specific projects and developments. Careful planning and implementation of water quality studies is necessary in optimizing the gathered information relative to study expenditures. As greater attention is directed toward river water quality both in the United States and throughout the developed and developing world, it is anticipated that the needs for planning and conduction of monitoring programs will increase. The purpose of this book is to present practical information on the planning and conduction of river water quality monitoring studies.

Chapter 1 presents some brief information on water pollution and quality, and summarizes typical limitations of water quality study plans. Chapter 2 presents an overview of principles for planning and conducting water quality studies, including the delineation of study objectives. Chapter 3 describes preliminary field work needed for a study, including general field reconnaissance and the assemblage of information on water uses, wastewater sources, and flow balances. Detailed information on the rationale for selecting hydraulic, water quality, and biological parameters is presented in Chapter 4. Chapter 5 addresses the factors for consideration in locating sampling stations and determining the appropriate frequency of sampling. Sampling equipment, sample collection, and planning for laboratory analyses are covered in Chapter 6. Finally, Chapter 7 contains information on data analysis and presentation, including the preparation of reports on water quality studies. Detailed technical information and annotated bibliographies are contained in ten appendices.

The author wishes to express his appreciation to several persons instrumental in the assemblage of this book. First, Ing. Juan Carlos Sanchez and Ing. Kathy Octavio of the Institute Technologico Venezolano del Petroleo (Intevep) in Los Teques, Venezuela were involved with the author in planning a river water quality monitoring program. Second, Debby Fairchild of the Environmental and Ground Water Institute at the University of Oklahoma conducted the computer-based literature searches basic to several appendices. Finally, and most important, the author is indebted to Ms. Leslie Rard of the Environmental and Ground Water Institute for her typing skills and dedication to the preparation of this manuscript.

The author also wishes to express his appreciation to the University of Oklahoma College of Engineering for its basic support of faculty writing endeavors, and to his family for their understanding and patience.

Larry W. Canter
Sun Company Professor
  of Ground Water Hydrology

January, 1985

To Donna, Doug, Steve, and Greg

CONTENTS

Chapter

LIST OF TABLES

LIST OF FIGURES

CHAPTER 1

INTRODUCTION

River water quality studies are often required for establishing baseline conditions, setting quality criteria and standards, monitoring temporal changes, and determining the impacts of specific projects and developments. Water quality impacts can occur from many types of projects, including the construction and operation of multi-purpose reservoirs, industrial complexes, and housing developments. Water pollution impacts may be visible in terms of floating debris and the occurrence of dead fish. Careful planning and implementation of water quality studies is necessary in optimizing the gathered information relative to study expenditures. The purpose of this book is to present practical information on the planning and conduction of water quality studies. The book does not contain a "cookbook" approach; however, it summarizes relevant concerns so that the user could apply the information in planning and conducting specific studies. Chapter 1 presents some brief information on water pollution and quality, typical limitations of water quality study plans, and the organization of the book to overcome these limitations.

BASIC INFORMATION ON WATER POLLUTION AND QUALITY

Water pollution can be defined in a number of ways; however, the basic features of most definitions address excessive concentrations of particular substances for sufficient periods of time to cause identifiable effects. Water quality represents a term associated with the composite analysis of physical, chemical, and bacteriological parameters. Physical parameters include color, odor, temperature, solids (residues), turbidity, oil, and grease. Subparameters for solids include suspended and dissolved solids as well as organic (volatile) and inorganic (fixed) fractions. Chemical parameters associated with the organic content of water include biochemical oxygen demand (BOD), chemical oxygen demand (COD), total organic carbon (TOC), and total oxygen demand (TOD). Inorganic chemical parameters include salinity, hardness, pH, acidity, alkalinity, iron, manganese, chlorides, sulfates, sulfides, havy metals (mercury, lead, chromium, copper, and zinc), nitrogen (organic, ammonia, nitrite, and nitrate), and phosphorus. Bacteriological parameters include coliforms, fecal coliforms, specific pathogens, and viruses.

In evaluating water pollution impacts associated with the construction and operation of a potential project, two main sources of water pollutants should be considered. Nonpoint pollutants refer to those substances which can be introduced into receiving waters as a result of urban or rural runoff. Point sources are related to specific discharges from municipalities or industrial complexes.

The effects of pollution sources on receiving water quality are manifold and dependent upon the type and concentration of pollutants. Soluble organics, as represented by high BOD wastes, cause depletion of oxygen. Trace quantities of certain organics cause undesirable taste and odors, and some may be biomagnified in the aquatic food chain. Suspended solids decrease water clarity and hinder photosynthetic processes; if solids settle and form sludge deposits, changes in benthic ecosystems result.   Color, turbidity, oils, and floating materials are of concern due to their aesthetic undesirability and possible influence on water clarity and photosynthetic processes.   Excessive nitrogen and phosphorus can lead to algal overgrowth with concomitant water treatment problems resulting from algae decay and interference with treatment processes.   Chlorides cause a salty taste to be imparted to water, and in sufficient concentration limitations on water usage can occur.   Acids, alkalies, and toxic substances have the potential for causing fish kills and creating other imbalances in stream ecosystems.   Thermal discharges can also cause imbalances as well as reductions in stream waste assimilative capacity. Stratified flows from thermal discharges minimize normal mixing patterns in receiving streams and reservoirs (Canter, 1977).

TYPICAL LIMITATIONS OF WATER QUALITY STUDY PLANS

Plans for the conduction of water quality studies are often incomplete, lacking background information and integration.   Typical limitations of plans are characterized by the following comments:

1. There is no overview section which describes the elements of the sampling plan and their relationship to each other.

2. An essential element in the planning of water quality studies is the careful delineation of study objectives.   There is only minimal information in the sampling plan which addresses objectives.

3. Background information on the study river is needed, with this information including a description of the drainage basin, river flows, water uses, wastewater discharges, and pertinent water quality standards or criteria.

4. Minimal back-up information is provided on sampling station locations, sampling frequencies, types of water quality analyses, and sediment and biota sampling.

5. The sampling plan does not address in detail the actual collection of samples and their identification.

6. The sampling plan includes a brief description of the analytical laboratory.   Additional information on laboratories, particularly as related to laboratory design and the development and use of one or more mobile laboratories, should be included.

7. Information on standard analytical procedures should be summarized, and a laboratory quality assurance program should be planned.

8. There is a brief discussion of the form to be used for data recording. This is an important issue, and careful detail needs to be provided in terms of subsequent data organization as well as analysis.

9. An important aspect of a river water quality study is the preparation of a report describing the study findings.   The study plan

should address this subject in a general manner, even though report details would have to be worked out in the future.

10.   Approaches for summarizing large amounts of information should be described, including the use of both water quality and aquatic biota or habitat indices.

ORGANIZATION OF BOOK

In order to provide positive suggestion to overcome the typical limitations of water quality study plans, this book is organized into six substantive chapters and ten technical appendices. Chapter 2 presents an overview of principles for planning and conducting water quality studies, including the delineation of study objectives.   Chapter 3 describes preliminary field work needed for a study, including general field reconnaissance and the assemblage of information on water uses, wastewater sources, and flow balances.

Detailed information on the rationale for selecting hydraulic, water quality, and biological parameters is presented in Chapter 4.   Chapter 5 addresses the factors for consideration in locating sampling stations and determining the appropriate frequency of sampling.   Sampling equipment, sample collection, and planning for laboratory analyses are covered in Chapter 6.   Finally, Chapter 7 contains information on data analysis and presentation, including the preparation of reports on water quality studies.

Detailed technical information is contained in ten appendices. Appendix A has a discussion of 12 steps associated with designing a monitoring network. Appendix C highlights the key features of a quality assurance program, and Appendix H contains supplementary information to Chapter 5 on the location of water quality monitoring stations. The remaining appendices include annotated bibliographies on the following topics:

Appendix B--general monitoring planning considerations

Appendix D--case studies of water quality monitoring

Appendix E--analytical techniques for water quality parameters

Appendix F--bacteriological water quality monitoring

Appendix G--biological monitoring

Appendix I--automatic sampling and remote sensing

Appendix J--water quality and biological indices

SELECTED REFERENCE

Canter, L.W., Environmental Impact Assessment, 1977, McGraw-Hill Book Company, New York, New York, Ch. 5.

CHAPTER 2

PRINCIPLES FOR PLANNING AND CONDUCTING
WATER QUALITY STUDIES

Several basic principles can be used for the comprehensive planning and conduction of a water quality study. An important perspective is the overall framework of the monitoring system as well as the individual components. Attention needs to be given to the collection of representative samples, analytical quality control, and delineation of study objectives. This chapter addresses these principles.

GENERAL FRAMEWORK FOR WATER QUALITY STUDIES

The most effectively planned water quality study is one which recognizes the multitude of elements within the study, even though not all of the elements are addressed during the initial stages. To illustrate this concept, Figure 1 provides a systems analysis framework for the planning and conduction of a water quality study (NUS Corporation, 1970). Initial review suggests that Figure 1 is overly complicated; however, careful examination of the individual elements reveals certain key components, with these being as follows:

(1)   Identify and describe bodies of water in each basin.

(2)   Identify and quantify water quality and environmental parameters for specific areas of each basin.

(3)   Identify and quantify present problem conditions and sources of pollution.

(4)   Identify and quantify potential future pollution problems and sources.

(5)   Identify and quantify water quality standards.

(6)   Identify the various surveillance agencies.

(7)   Define the objectives of the present (or near-term) surveillance system.

(8)   Identify alternative sample collection techniques and equipment along with associated facilities, personnel, and other resources required for operation and support.

(9)   Specify the desirable number and approximate locations of additional water quality surveillance stations (based on the assumption that some are existing).

(10) Specify collection requirements of the surveillance stations.

(11) Define alternative analytical systems compatible with the collection

5

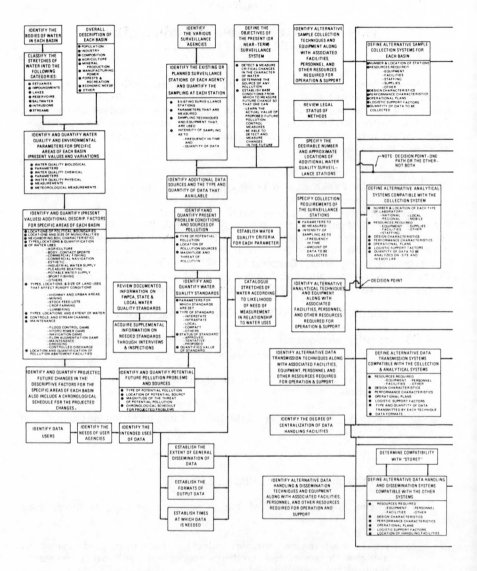

Figure 1: General Systems Analysis Framework (NUS Corporation, 1970)

system.

(12)   Define alternative data transmission systems compatible with the collection and analytical systems.

End products of the systems analysis framework shown in Figure 1 would be a definition of the selected collection system, the selected analytical

Figure 1: (continued)

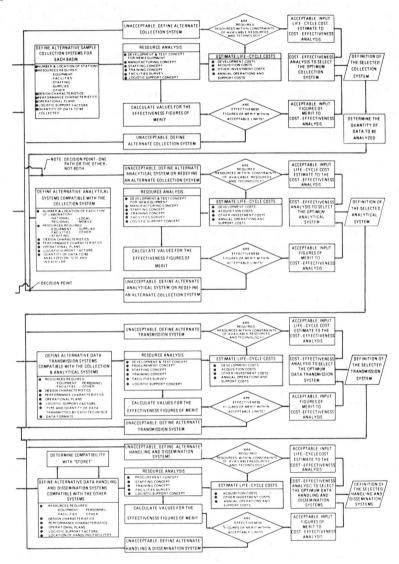

system, the selected transmission system, and the selected data handling and dissemination systems.

Complete design of a water quality sampling network involves a series of steps. Table 1 summarizes 12 steps which can be used for the design of a sampling network (Sanders, 1980). The 12 steps listed should be viewed as general guidance as opposed to specific rules for design. Detailed information related to each of these steps is in Appendix A. Steps 1 through 3 are related to defining the objectives of the sampling program. Steps 4 through 7 are self-explanatory, while Step 8 involves revisions of sampling station locations and sampling frequencies based on information gathered to that point. Step 9 on the development of operational plans includes consideration of sampling equipment, field or in situ monitoring, as well as laboratory analyses. Step 10 focuses on data formats and the associated data reporting. Step 11 relates to adjustment of the first 10 steps for study compatibility. Finally, Step 12 is associated with the preparation of a network design report.

Table 1: Steps Related to Sampling Network Design (Sanders, 1980)

1. Determine monitoring objectives and relative importance of each.

2. Express objectives in statistical terms.

3. Determine budget available for monitoring and amount to be allocated for each objective.

4. Define the characteristics of the area in which the monitoring is to take place.

5. Determine water quality variables to be monitored.

6. Determine sampling station locations.

7. Determine sampling frequency.

8. Compromise previous objective design results with sub-jective considerations.

9. Develop operating plans and procedures to implement the network design.

10. Develop data and information reporting formats and procedures.

11. Develop feedback mechanisms to fine tune the network design.

12. Prepare a network design report.

A number of additional monitoring planning considerations are listed in Table 2, with the cited author(s) and abstracts included, in order, in Appendix B.

REPRESENTATIVE SAMPLING

One critical issue in planning a water quality study is to insure that representative samples are collected. The term "representative" means that the sample resembles the population of all possible samples in some way (U.S. Geological Survey, 1977). To achieve representativeness for data comparability and consistency, it is necessary to minimize, or at least standardize, sampling bias as it relates to site selection, sampling frequency, sample collection,

Table 2: References on General Monitoring Planning Considerations

Authors (Year)	Comments
Beach and Beach (1977)	Techniques of sampling for wastewater discharges.
Chakrabarti et al. (1978)	Evaluation of preservation techniques for some anionic species in water samples.
Casey and Walker (1981)	Possibilities for in situ measurements for water chemistry and sample storage and filtration requirements.
Collins (1972)	Features of portable water analysis kit.
Drake (1975)	Programmer's manual for data handling system for water quality surveys.
Fenlon and Young (1982)	Objectives of program for chemical surveillance of rivers.
Heidtke and Armstrong (1979)	Cost effective sampling strategy for detection of stream standard violations.
Landwehr (1978)	Properties of geometric mean in relation to water quality standards.
May (1981)	Instructional module for collecting stream samples for water quality.
Morris and Ames (1977)	Implementation of an automation system for water quality laboratories.
National Field Investigations Center (1974)	Comparison of grab and automatic sampling techniques.
Otson et al. (1979)	Effects of sampling, shipping, and storage on total organic carbon levels in water samples.
Reckhow (1980)	Technique for presenting water quality sampling data.
Sanders and Adrian (1978)	Sampling frequency as a function of river flow characteristics.
Schock and Schock (1982)	Effects of sampling container type on pH and alkalinity.
Shnider and Shapiro (1976)	Evaluation of efficiency of surface water quality monitoring networks.

Table 2: (continued)

Authors (Year)	Comments
Tuschall, Viborsky and Surles (1977)	Use of fluorescent tracers for continuous flow monitoring.
Williams (1979)	Preservation methods for nitrates in water samples.

sampling devices, and sample handling, preservation, and identification. Therefore, in order to best achieve representative sampling, it is necessary to use a consistent approach in sample site selection, frequency of sampling, sampling equipment, and other associated sampling activities.

ANALYTICAL QUALITY CONTROL

Due to the large number of water quality and sediment parameters and biota testing to be done in a baseline study, it is necessary that appropriate analytical quality control be achieved for the planned laboratory analyses. The continued output of high quality, reliable data depends not only on the use of acceptable analytical methods by competent and conscientious analysts, but also on a prescribed program of regular laboratory quality control (U.S. Geological Survey, 1977). One of the things which should be considered for a sampling program is the development of a quality assurance plan. This concept has been used in the United States by the Environmental Protection Agency, and it represents an organized program to insure appropriate quality control for analytical testing. Appendix C contains information on quality assurance program planning.

DELINEATION OF OBJECTIVES

An important element in planning a water quality study is the careful delineation of the objectives. The objectives should be put into writing for several reasons (Kittrell, 1969). One reason is that the act of trying to write objectives will require careful consideration of what the objectives actually should be. A written objective is less likely to be misunderstood by the study participants than if the participants are trying to deal with verbal statements. Written objectives will also provide an effective means of communication with the study sponsors, and indicate that a systematic and careful plan has been developed. Written objectives will also enable an evaluation of the study at various times in terms of whether or not the objectives are being met. In that sense, the objectives serve as a basis for study evaluation. Examples of objectives for determining water quality at a single point are shown in Table 3, and water quality at a series of related points or stations are shown in Table 4 (Kittrell, 1969).

CASE STUDIES

In planning a water quality monitoring program it is often useful to review extant monitoring programs. Table 5 identifies five case studies, with their abstracts contained in Appendix D.

Table 3: Objectives for Studying Water Quality at a Single Point (Kittrell, 1969)

---

1. Establishment of a base-line record of water quality.

2. Investigation of suitability as a source of industrial water supply.

3. Day-to-day monitoring of raw water sources of industrial water supply.

4. Monitoring effects of waste discharges.

5. Surveillance to detect adherence to or violation of water quality standards.

6. Detection of sudden changes in water quality caused by slugs of wastes resulting from spills, deliberate discharges or treatment plant failures.

Table 4: Objectives for Studying Water Quality at Related Points (Kittrell, 1969)

---

1. Determination of patterns of pollution downstream from waste discharges and effects on water uses.

2. Determination of adherence to or violation of water quality standards.

3. Determination of characteristics and rates of natural purification of streams.

4. Projection of effects of pollution to other conditions of flow and temperature than those occurring during study.

5. Estimation of waste assimilative capacities of streams.

6. Estimation of reductions in waste loads necessary to meet water quality requirements.

7. Determination of existing water quality before some change in conditions, such as a new or increased waste discharge.

Table 5: References on Case Studies of Water Quality Monitoring

Authors (Year)	Comments
Burge, Herrmann and Matthews (1979)	Use of water and meteorological data collection stations in the Great Smoky Mountains National Park.
Raytheon Company (1982)	Description of environmental monitoring program at the Oak Ridge National Laboratory.
Shackelford and Keith (1976)	List of organic compounds identified in water samples.
U.S. Environmental Protection Agency (1975)	Description of model state water monitoring program.
U.S. Environmental Protection Agency (1981)	Review of Federal and state water monitoring activities in EPA Region V.

SELECTED REFERENCES

Beach, M.I. and Beach, J.S., Jr., "Sample Acquisition the First Step in Water Quality Monitoring", International Workshop on Instrumentation and Control for Water and Wastewater Treatment and Transport Systems, 1977, N-CON Systems Co., London, England.

Burge, R.E., Herrmann, R. and Matthews, R.C., Jr., "Remote Sensing of Water Quality and Weather in Great Smoky Mountains National Park: Phase I", R/RM-30, 1979, U.S. Forest Service, Atlanta, Georgia.

Chakrabarti, C.L. et al., "Preservation of Some Anionic Species in Natural Waters", American Water Works Association Journal, Vol. 70, No. 10, Oct. 1978, pp. 560-565.

Casey, H. and Walker, S.M., "Storage and Filtration of Water Samples", International Environment and Safety, 1981, pp. 16-17.

Collins, W.H., "Improved Water Analysis Kit", FIRL-F3222-02/03, Dec. 1972, Franklin Institute Research Laboratories, Philadelphia, Pennsylvania.

Drake, T.L., "A Systems Analysis of Water Quality Survey Design. Appendix VIII. Documentation Data Handling System Programmer's Manual", AD-A036 521, 1975, Clemson University, Clemson, South Carolina.

Fenlon, J.S. and Young, D.D., "Chemical Surveillance of Rivers", Water Pollution Control, Vol. 81, No. 3, 1982, pp. 343-357.

Heidtke, T.M. and Armstrong, J.M., "Probabilistic Sampling Model for Water Quality Management", Journal of the Water Pollution Control Federation, Vol. 51, No. 12, Dec. 1979, pp. 2916-2927.

Kittrell, F.W., "A Practical Guide to Water Quality Studies of Streams", CWR-5, 1969, 135 pp., Federal Water Pollution Control Administration, U.S. Department of Interior, Washington, D.C.

Landwehr, J.M., "Some Properties of the Geometric Mean and Its Use in Water Quality Standards", Water Resources Research, Vol. 14, No. 3, June 1978, pp. 467-473.

May, F.C., "Vocational Education Training in Environmental Health Sciences: Collecting Stream Samples for Water Quality, Module 16", DE/OVAE-80-0088-16, 1981, Rockville, Maryland.

Morris, W.F. and Ames, H.S., "Automation of the National Water Quality Laboratories, U.S. Geological Survey. I. Description of Laboratory Functions and Definition of the Automation Project", 1977, Lawrence Livermore Laboratory, Livermore, California.

National Field Investigations Center, "Comparison of Manual (GRAB) and Vacuum Type Automatic Sampling Techniques on an Individual and Composite Sample Basis", EPA-330/1-74-001, 1974, Denver, Colorado.

NUS Corporation, "Design of Water Quality Surveillance Systems, Phase I--Systems Analysis Framework", 16090 DBJ 08/70, Aug. 1970, 303 pp., Pittsburgh, Pennsylvania (report prepared for Federal Water Quality Administration, U.S. Department of Interior, Washington, D.C.).

Otson, R. et al., "Effects of Sampling, Shipping, and Storage on Total Organic Carbon Levels in Water Samples", Bulletin of Environmental Contamination and Toxicology, Vol. 24, No. 3, Oct. 1979, pp. 311-318.

Raytheon Company, "Environmental Monitoring Report, United States Department of Energy, Oak Ridge Facilities", Y/UB-16, 1982, Washington, D.C.

Reckhow, K.H., "Techniques for Exploring and Presenting Data Applied to Lake Phosphorus Concentration", Canada Fisheries Research Board Journal, Vol. 37, No. 2, Feb. 1980, pp. 290-294.

Sanders, T.G., editor, "Principles of Network Design for Water Quality Monitoring", July 1980, 312 pp., Colorado State University, Ft. Collins, Colorado.

Sanders, T.G. and Adrian, D.D., "Sampling Frequency for River Quality Monitoring", Water Resources Research, Vol. 14, No. 4, Aug. 1978, pp. 569-576.

Schock, M.R. and Schock, S.C., "Effect of Container Type on pH and Alkalinity Stability", Water Resources, Vol. 16, No. 10, 1982, pp. 1455-1464.

Shackelford, W.M. and Keith, L.H., "Frequency of Organic Compounds Identified in Water", EPA/600/4-76/062, 1976, U.S. Environmental Protection Agency, Athens, Georgia.

Shnider, R.W. and Shapiro, E.S., "Procedures for Evaluating Operations of Water Monitoring Networks", EPA/600/4-76/050, 1976, U.S. Environmental Protection Agency, Las Vegas, Nevada.

Tuschall, J.R., Jr., Viborsky, R. and Surles, T., "Continuous Flow Monitoring Using Fluorescent Tracers", Pollution Engineering, Vol. 9, No. 10, Oct. 1977, p. 59.

U.S. Environmental Protection Agency, "Model State Water Monitoring Program", EPA/440/9-74/002, 1975, Washington, D.C.

U.S. Environmental Protection Agency, "Assessment of the Region V Water Monitoring Activities", EPA-905/4-81-001, 1981, Chicago, Illinois.

U.S. Geological Survey, "National Handbook of Recommended Methods for Water-Data Acquisition", Ch. 5--Chemical and Physical Quality of Water and Sediment, Jan. 1977, pp. 5-7 through 5-16c, Reston, Virginia.

Williams, T.J., "An Evaluation of the Need for Preserving Potable Water Samples for Nitrate Testing", Journal of the American Water Works Association, Vol. 71, No. 3, Mar. 1979, pp. 157-160.

CHAPTER 3

PRELIMINARY FIELD WORK

An effective water quality study requires preliminary field work and data gathering exercises. This includes field reconnaissance as well as the development of a preliminary flow balance for the system to be monitored. Additionally, information is needed on water uses from, as well as wastewater discharges into, the stream to be monitored. This chapter addresses each of these topics.

FIELD RECONNAISSANCE

Field reconnaissance of the pertinent study area is one of the most important phases of the baseline data gathering effort. Reconnaissance information can be used for the development of more detailed plans for the sampling program. A thorough reconnaissance provides the basis for the soundest possible study plan, insures smoother operation of the study, and reduces confusion, time, effort and cost in the long run (Kittrell, 1969). Elements for consideration in the field reconnaissance program include the reconnaissance crew, preliminary tour, studies of water flow and stream characteristics, supplies and services necessary for sampling, and the development of contacts with local officials who may be of assistance in the sampling program.

Reconnaissance Crew

There are no fixed rules relative to the size of a reconnaissance crew. A minimum crew should consist of a stream biologist, environmental engineer, and chemist. It is particularly important that a biologist be a member of the reconnaissance team since visual observations and preliminary examinations of bottom organisms can be helpful in revealing the historical record of pollution severity at a given location. Preliminary findings by the biologist should influence the final planning of the study. The environmental engineer would be most interested in stream flows and characteristics as well as water uses and wastewater discharges. The latter two items will be addressed in subsequent sections. The chemist should be oriented to sample collection and water and sediment analyses.

Preliminary Tour

A preliminary tour of the study area can be invaluable to the reconnaissance team in enabling them to get the general "lay of the land" and identify various basin characteristics. This preliminary tour could consist of an aerial flight over the study area coupled with visits to strategic towns along the river. Additionally, a preliminary boat tour of the river can also be beneficial in that both water withdrawal and wastewater discharge points can be identified along with unique flow characteristics such as the presence of sand bars, rapids, tributary streams, and other information of this type.

Time of Water Travel and Stream Characteristics

One of the major purposes of a preliminary field reconnaissance will be the conduction of some flow studies as well as the identification of certain stream characteristics. Studies related to river flows can be invaluable prior to final selection of sampling station locations. Flow studies will be addressed in Chapter 4.

Another important activity is to determine access to the study river via bridges and roads. This is an important consideration in establishing sample station locations and in providing appropriate benchmarks for consistency in sampling locations. Detailed notes of any observations of stream characteristics should be made promptly. Notes should include general impressions of depths, currents, bends, widths, types of bottom, and sensory evidences of pollution, such as excessive plankton or attached growth, floating materials, oil, color, suspended matter, sludge deposits, gas bubbles, and odor (Kittrell, 1969). Special attention should be paid to the areas that may have been tentatively selected for sampling station locations.

Supplies and Services

Sources of needed supplies should be located as a part of the field reconnaissance. Supplies may include ice, distilled water, hardware, laboratory reagents, and minor equipment. This does not have to be done in detail, but thought does need to be given to the availability of these supplies prior to the initiation of the water quality study. Additionally, consideration needs to be given to the availability of repair services, such as automotive, outboard motor, and electrical.

Local Assistance

In many instances it will be necessary to have persons who can be contacted for providing local assistance during the sampling program. Contact should be made with persons in local municipalities that could provide assistance during the field sampling program. It would be desirable to have one person living in the vicinity of each sampling station who could render assistance on an as needed basis during the sampling program.

In summary, a preliminary field reconnaissance is an important element in the development of a sampling plan for a river. It would be effort well spent to conduct a field reconnaissance program prior to finalization of the sampling plan for a river.

USES OF RIVER WATER

The uses of water from the study river, either present or future, constitute one of the prime reasons for the planning and conduction of a study of water quality and contamination. If streams were used only for waste disposal, then there would be minimal need for the conduction of water quality studies. One of the basic items needed in a comprehensive study of a river is information on water usage. This information is necessary due to its importance relative to river flows and requirements for various levels of quality depending upon the water usage.

To serve as an illustration of the quality requirements depending upon water usage, Table 6 summarizes the requirements for public and private water supplies utilized by the State of Oklahoma (Oklahoma Water Resources Board, 1979).

Table 6:   Stream Quality Requirements for Public and Private Drinking Water Supplies (Oklahoma Water Resources Board, 1979)

Characteristics	Standards
Physical	
Color	Color producing substances from other than natural sources, shall be limited to concentrations equivalent to 75 color units (CU).
Odor	Taste and odor producing substances from other than natural origin shall be limited to concentrations that will not interfere with the production of a potable water supply by modern treatment methods.
Temperature	At no time shall heat be added to any stream in excess of the amount that will raise the temperature of the receiving water more than 5°F. In streams, temperature determinations shall be made by averaging representative temperature measurements of the cross sectional area of streams at the end of the mixing zone. The normal daily and seasonal variations that were present before the addition of heat from other than natural sources shall be maintained. The maximum temperature due to man made causes shall not exceed 68°F in trout streams, 84°F in smallmouth bass streams, or 90°F in all other streams and lakes.
Inorganic Elements (mg/l)	
Arsenic	0.05
Barium	1.0

Table 6: (continued)

Characteristics	Standards
Cadmium	0.01
Chromium	0.05
Copper	1.0
Fluoride (at 90°F)	1.6
Lead	0.05
Mercury	0.002
Nitrates	10.0
pH	The pH value shall be between 6.5 and 8.5, pH values less than 6.5 or greater than 8.5 must not be due to water discharge(s).
Selenium	0.01
Silver	0.05
Zinc	5.0
Organic Chemicals (mg/l)	
Cyanide	0.2
Detergents (total)	0.2
Methylene blue active substances	0.5
Oil and grease	All waters shall be maintained free of oil and grease to prevent a visible film of oil or globules of oil or grease on or in the water. Oil and grease shall not be present in quantities that adhere to stream banks and coat bottoms of water courses or which cause deleterious effects to the biota. For public and private water supplies, the water shall be maintained free from oil and grease and taste and odors that emanate from petroleum products.
Phthalate esters	0.003

Table 6: (continued)

Characteristics	Standards

Microbiological

Coliform organisms

The bacteria of the fecal coliform group shall not exceed a monthly geometric mean of 200/100 ml, as determined by multiple tube fermentation or membrane filter procedures based on a minimum of not less than five (5) samples taken over not more than a thirty (30) day period. Further, in no more than 10% of the total samples during any thirty (30) day period shall the bacteria of the fecal coliform group exceed 400/100 ml.

Table 7 summarizes information on quality requirements for various industrial uses of water (Canter, 1977). Table 7 is not intended to address all possible industrial uses in a river basin, but it does show that various uses require different levels of quality. In addition to water usage for public and private water supplies as well as industries, Table 8 lists several other categories of water use (Kittrell, 1969). Standards or criteria applicable to water usage for agriculture, recreation, and other uses as shown in Table 8 can be found in an excellent reference source by McKee and Wolf (1963).

The water uses listed as 1 through 3 in Table 8 involve the withdrawal of water from specific points in the study river. It would be desirable to have information on the quantities of water withdrawn to supply municipal, industrial, and agricultural needs. Some information would probably be available from the municipalities and industries involved with withdrawal of water from the study river. If adequate information is not available, estimations could be used for developing a preliminary flow balance for the river. Judgment will need to be exercised at this point, with primary attention devoted to the major water uses within the basin. In addition to the quantities of water withdrawn for meeting the various needs, additional information which would be useful includes the number of persons served by municipal supplies, the quantity or value of goods manufactured with industrial supplies, and the acres irrigated or the value of crops produced by agricultural supplies (Kittrell, 1969).

Useful data on other uses listed in Table 8 include the number of swimmers, the value of bathing facilities, the number of boat licenses, estimates of the total numbers of boats and their value, numbers of fish caught by sport fishermen and commercial fishermen, and numbers of aquatic waterfowl hunted in the area. Navigation use data includes tons of materials or ton-miles transported, as well as the total values of cargos being shipped to and from port areas.

Table 7:   Summary of Specific Quality Characteristics of Surface Waters That Have Been Used as Sources for Industrial Water Supplies[a] (Canter, 1977)

| Characteristic | Boiler makeup water | | Cooling water | | | | Process water | | | | | |
| | Industrial, 0-1,500 psig | Utility, 700-5,000 psig | Fresh | | Brackish[b] | | Textile industry, SIC-22 | Lumber industry, SIC-24 | Pulp and paper industry, SIC-26 | Chemical industry, SIC-28 | Petroleum industry, SIC-29 | Primary metals industry, SIC-33 |
			Once through	Makeup recycle	Once through	Makeup recycle						
Silica (SiO2)	150	150	50	150	25	25	–	–	50	–	50	–
Aluminum (Al)	3	3	3	3	–	–	–	–	–	–	–	–
Iron (Fe)	80	80	14	80	1.0	1.0	0.3	–	2 6	5	15	–
Manganese (Mn)	10	10	2.5	10	0.02	0.02	1.0	–	–	2	–	–
Copper (Cu)	–	–	–	–	–	–	0.5	–	–	–	–	–
Calcium (Ca)	–	–	500	500	1,200	1,200	–	–	–	200	220	–
Magnesium (Mg)	–	–	–	–	–	–	–	–	–	100	85	–
Sodium and potassium (Na + K)	–	–	–	–	–	–	–	–	–	–	230	–
Ammonia (NH3)	–	–	–	–	–	–	–	–	–	–	–	–
Bicarbonate (HCO3)	600	600	600	600	180	180	–	–	–	600	480	–
Sulfate (SO4)	1,400	1,400	680	680	2,700	2,700	–	–	–	850	570	–
Chloride (Cl)	19,000	19,000	600	500	22,000	22,000	–	–	200[c]	500	1,600	500
Fluoride (F)	–	–	–	–	–	–	–	–	–	–	1.2	–
Nitrate (NO3)	–	–	30	30	–	–	–	–	–	–	8	–
Phosphate (PO4)	–	50	4	5	5	5	–	–	–	–	–	–
Dissolved solids	35,000	35,000	1,000	1,000	35,000	35,000	150	–	1,080	2,500	3,500	1,500
Suspended solids	15,000	15,000	5,000	15,000	250	250	1,000	d	–	10,000	5,000	3,000
Hardness (CaCO3)	5,000	5,000	850	850	7,000	7,000	120	–	475	1,000	900	1,000
Alkalinity (CaCO3)	500	500	500	500	150	150	–	–	–	500	–	200
Acidity (CaCO3)	1,000	1,000	0	200	0	0	–	–	–	–	–	75
pH, units	–	–	5.0-8.9	3.5-9.1	5.0-8.4	5.0-8.4	6.0-8.0	5-9	4.6-9.4	5.5-9.0	6.0-9.0	3-9
Color, units	1,200	1,200	–	1,200	–	–	–	–	360	500	25	–
Organics												
Methylene blue active substances	2[e]	10	1.3	1.3	–	1.3	–	–	–	–	–	–
Carbon tetrachloride extract	100	100	f	100	f	100	–	–	–	–	–	30
Chemical oxygen demand (O2)	100	500	–	100	–	200	–	–	–	–	–	–
Hydrogen sulfide (H2S)	–	–	–	–	4	4	–	–	–	–	–	–
Temperature, °F	120	120	100	120	100	120	–	–	95[g]	–	–	100

[a] Unless otherwise indicated, units are mg/liter and values are maximums. No one water will have all the maximum values shown.
[b] Water containing in excess of 1,000 mg/liter dissolved solids.
[c] May be ≤ 1,000 for mechanical pulping operations.
[d] No large particles ≤ 3 mm diameter.
[e] 1 mg/liter for pressures up to 700 psig.
[f] No floating oil.
[g] Applies to bleached chemical pulp and paper only.

Table 8:   Categories of Water Use (Kittrell, 1969)

1. Municipal (public) water supply.

2. Industrial water supply.

3. Agricultural water supply.

    a. Domestic (private) farm supply
    b. Irrigation
    c. Livestock watering

4. Recreation.

    a. General
    b. Swimming, wading, skiing
    c. Boating
    d. Esthetic enjoyment

Table 8: (continued)

_____

5. Propagation of fish and other aquatic life and wildlife.

    a. Sport fishing
    b. Commercial fishing
    c. Fur trapping

6. Hydropower production.

7. Navigation.

8. Waste disposal.

    a. Low flow augmentation

_____

WASTEWATER SOURCES

Wastewater discharges into the study river may include those from municipalities and industries. These discharges are commonly designated as point sources. In addition, consideration needs to be given to nonpoint sources as characterized by urban runoff and agricultural drainage. Information should be gathered on the types and quantities of wastewater discharges into the study river. With this information, estimates could then be made of the total wastewater loadings in the river based on published information on wastewater characteristics by source type. Additional information which would be useful include the types of wastewater treatment processes serving municipalities and industries. Since the effort for gathering pertinent information on wastewater discharges can be extensive, attention should first be devoted to major sources, with approximations used for lesser significant sources.

PRELIMINARY FLOW BALANCE

A useful exercise in the early stages of planning for the study of a river will be the development of a preliminary balance for river flow. To achieve this balance it will be necessary to have flow information from the river as well as estimates for water withdrawals for various uses and wastewater discharges from both point and nonpoint sources. One reason for developing a preliminary flow balance is to aid in detailed planning of the study as well as selection of sampling station locations. An additional purpose is to determine if there are any significant flow increases or losses resulting from interchange of river water with underlying aquifers. This can be a significant factor in water quality studies, and its relevance to the river basin under study should be determined. A more detailed discussion of the hydrological measurements necessary in a comprehensive study is contained in Chapter 4.

SELECTED REFERENCES

Canter, L.W., Environmental Impact Assessment, 1977, McGraw-Hill Book Company, New York, New York, Ch. 5.

Kittrell, F.W., "A Practical Guide to Water Quality Studies of Streams", CWR-5, 1969, 135 pp., Federal Water Pollution Control Administration, U.S. Department of Interior, Washington, D.C.

McKee, J.E. and Wolf, H.W., "Water Quality Criteria", Pub. No. 3-A, Second Edition, 1963, 548 pp., State Water Quality Control Board, Sacramento, California.

Oklahoma Water Resources Board, "Oklahoma's Water Quality Standards--1979", Pub. No. 101, 1979, Oklahoma City, Oklahoma.

CHAPTER 4

SELECTION OF PARAMETERS TO BE MONITORED

An important element in planning a river study is to carefully identify hydraulic, water quality, biological, and sediment parameters to be monitored. This section addresses the factors for consideration in selection of these parameters.

HYDRAULIC PARAMETERS

Due to variations of the concentrations of various water quality constituents with time and stream flow, it is necessary that information be available on flow frequencies in the study river. In addition, an important factor in selecting sampling station locations is the extent of mixing within the river.

Stream Flow Data

Stream flow can be estimated from precipitation data, but it is much better to use information based on actual stream flow measurements. Stream flow measurements include measurements of stream discharges as well as water surface elevations, with the collection of stream flow data being called stream-gaging (Butler, 1957). Most stream-gaging discharge measurements are made by the area-velocity method using a current meter to measure the velocity. In the area-velocity method the channel cross section is divided into subareas, each subarea is multiplied by its mean sub-areal velocity to determine the sub-areal discharge, and the sub-areal discharges are added to determine total stream discharge. The mean velocity is computed by dividing the total discharge by the total cross sectional area. Field data for an area-velocity discharge measurement consists of measurements of velocity, depth, and width-stationing at various positions in the channel cross section.

Velocity Measurements

There are several approaches for velocity measurement, including the use of surface float methods as well as dyes and current meters. A rough method for preliminary estimates of stream velocity consists of dropping sticks or other floatable objects from bridges in the current of the stream reach under observation, and noting the time required for them to float an estimated ten feet, or some other convenient distance. These velocity estimates tend to be very inaccurate for use in interpretation of data or final reporting, but they can be useful in preliminary planning of studies and in subsequent more precise measurements of stream velocity (Kittrell, 1969).

Surface floats may be followed downstream and timed for known distances to determine velocities. This requires the exercise of considerable judgment since floats tend to travel into quiet or eddy areas, or become stuck on tree limbs, the stream bank, or other obstacles. The surface water velocity as determined through the use of floats is greater than the average for the entire stream, and a correction factor must be applied. An average velocity of about 85 percent of that of the surface velocity is a reasonable rule-of-thumb value (Kittrell, 1969).

A fairly accurate method of measuring velocities involves following a tracer downstream. A common type of tracer are dyes, with an example being the use of Rhodamine WT. This dye can be detected in concentrations as low as 0.05 part per billion (ppb) by a fluorometer (Kittrell, 1969). The dye must be distributed across the stream at the upstream point as nearly instantaneously as possible. The ideal distribution produces a narrow band of tracer in uniform concentration across the stream. The band of tracer mixes with water ahead of and behind it by diffusion, or longitudinal mixing, as it moves downstream to produce an increasingly wider band. The peak concentration remains near, but somewhat downstream of, the center line of the band and decreases as longitudinal mixing proceeds. The times-of-water travel to downstream points are the differences between the time the dye was added to the stream and the time the centroid of the dye mass arrives at downstream points. Peak concentrations of Rhodamine WT dye at downstream points in the range of 1-10 ppb allow satisfactory definition of the downstream dye concentration curve.

Several methods have been developed for calculating the dosage of dye needed at the upstream point. The simplest method is calculation of the weight of the dye required to produce a concentration of 1 ppb in the estimated total volume of water in the reach between the two points where time-of-travel is being determined. The calculated dosage produces concentrations in excess of 1 ppb at the downstream point, since the dye is not actually mixed with the total volume of water in the reach. The stream should be sampled frequently as the dye arrives at the downstream point to define the curve of concentration versus time, with special emphasis given to the time of peak occurrence. The frequency of measurement may be varied from once each minute to once every 15 minutes, depending on how wide the band of dye has become at the sampling point. The dye may be missed altogether by overestimating the time required for it to travel downstream; on the other hand, time may be wasted in waiting for it to arrive if the time-of-travel is underestimated (Kittrell, 1969).

Perhaps the most common method of measuring stream velocity is by the use of current meters. The standard vertical-axis cup-vane Price current meter is commonly used in the United States and throughout much of the world. The meter is held in the water on a rod if the water is shallow enough for wading. In deep water, it is usually suspended on a cable and lowered into the water from a bridge or boat. The cup-vane assembly with a 5-inch outside diameter rotates very nearly in proportion to the velocity of the water. The rate of rotation is determined by stopwatch and is converted to point velocity in feet per second (fps) through application of an equation as follows: $V = 2.20 R + 0.30$, where R is meter speed in revolutions per second and V is velocity in feet per second (Butler, 1957).

For measuring velocities in streams with a depth of 0.5 feet or less, greater accuracy may be achieved by using a small Price-type meter called the pigmy current meter. The cup-vane assembly for the pigmy meter is about 2 inches in diameter. For the same water velocity the pigmy meter rotates faster than the standard Price-type meter, and this is illustrated by the following equation: $V = 0.975 R + 0.010$.

Depth Measurements

One of the questions which arises in conjunction with measuring the velocity at a point is the depth at which the velocity should be taken. Numerous tests have indicated that the mean velocity in the vertical section is closely represented by the average of the velocities occurring at points 20 and 80 percent of the depth below the water surface (Butler, 1957). Sometimes, however, the stream depths are so shallow that the 0.2-0.8 depth method of observation is not feasible. In this case, the following criteria can be used to determine the water depth at which the meter should be placed: (1) for water depths greater than 0.5 ft and using the standard Price-type meter, position the meter at 0.6 of the depth from the surface; (2) for water depths less than 0.5 ft and using the standard Price-type meter, position the meter at 0.5 of the depth from the surface; (3) for water depths greater than 0.35 ft and using the pigmy meter, position the meter at 0.6 of the depth from the water surface; and (4) for water depths less than 0.35 feet and using the pigmy meter, position the meter at 0.5 of the depth from the surface (Butler, 1957).

Area Measurements

Another thing which must be done in measuring stream flow by the area-velocity method is to subdivide the stream cross section into areas. The widths of the subdivisions and of the entire stream may be measured by means of a calibrated line as shown in Figures 2 and 3 (Butler, 1957). If the section chosen for the measurement is not at right angles to the direction of flow, the width data can be adjusted by multiplying by the cosine of the angle of measurement. If a stream can be waded, the required depth of flow at the various positions in the cross section can be determined by reading a calibrated rod placed in the stream.   Otherwise, the depths can be measured by lowering a weighted sounding line from a bridge, cable car, or boat. It may be necessary to correct for line drift in the case of high bridges or rapidly flowing streams (Butler, 1957).

Stream discharge may be computed from the field observations of depth, width, and velocity by either the mid-section method or the mean-section method.   Figure 2 shows a channel cross section illustrating the mid-section method. Dashed lines (a) through (f) represent observed depths. The mean-in-vertical velocity is determined for each depth line. The depth and mean-in-vertical velocity data for each line are assumed to be representative for a subarea extending halfway on each side toward the next observation line. Thus as shown by the crosshatching, the assumed channel cross section finally consists of a series of rectangular subareas, each assumed to be flowing in a uniform velocity equal to that at the depth-observation line. For example, the mean velocity for subarea 1-2-3-4 is assumed to be equal to the mean velocity for depth line e (Butler, 1957).

Figure 3 illustrates the mean-section method.   As shown by the cross-hatching, the channel cross section is assumed to consist of a series of trapezoidal subareas, each flowing at a uniform velocity equal to the mean of the velocities at the two bordering depth observation positions. For example, the mean velocity for the subarea between depth-observation lines (b) and (c) is assumed to be equal to $(V_b + V_c)/2$. The mean-section method is considered to be slightly more accurate, but the mid-section method is faster and is probably the better method for general use (Butler, 1957). Depending upon the size of the stream, the sub-areal divisions may exceed 20 in some locations. Figures 2 and 3 illustrate a much smaller number of subsections than this.

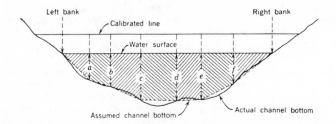

Figure 2: Channel Subdivision by the Mid-Section Method (Butler, 1957)

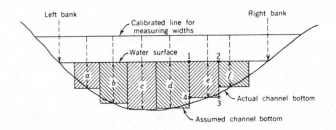

Figure 3: Channel Subdivision by the Mean-Section Method (Butler, 1957)

If possible, flow measurements should be made at a given location under at least three different flow conditions. This would allow for the development of more detailed information regarding flow patterns, and provide the opportunity of projecting flows to other conditions. The resulting travel times from one point to another plotted against the corresponding stream discharges for each stream section provides curves from which other travel times may be obtained by interpolation or extrapolation.

In some cases for small streams, or perhaps for larger streams under high flow conditions, it may not be possible to make flow measurements. If this is the case when a water quality survey is being made on a study river, it is possible to utilize an indirect method to determine the peak discharge. One of the empirical approaches which can be used involves the Manning equation (Butler, 1957). If the length of channel under study (commonly called a reach of channel) is of such length and roughness that channel friction head is appreciable, the Manning equation is applicable in the following form:

$$Q = \frac{1.49}{n} A_e R_e^{2/3} \left[\frac{hc}{L}\right]^{1/2}$$

where Q is the discharge in cubic feet per second, n is the Manning roughness factor and ranges from about 0.010 to about 0.100, A_e is the mean effective cross-sectional area in square feet for a given reach of channel, R_e is the corresponding mean effective hydraulic radius (area divided by the wetted perimeter), hc is the head loss in feet caused by friction along the bottom and walls of the channel, and L is the reach length in feet.   The terms A_e and R_e may be taken as the average of the two n values for A and R if the n values are equally representative; otherwise, A_e and R_e should be weighted in favor of the more representative values.   The term hc can be approximated as the difference in water elevation from the upstream to the downstream point.

Flow Analysis

Another hydraulic consideration which must be addressed in a water quality study is the frequency of flows. Of particular importance is the drought flow of a stream since this is used to determine the capacity of a river to assimilate waste materials.   A drought flow may be defined as the lowest average discharge which occurs for a defined time period (1, 7, 15 days) in a particular season in a base unit of time (1, 2, 10 years). It is customary to adopt a 7-day period as the drought duration, and 1 year as the base unit of time (IHD-WHO Working Group on the Quality of Water, 1978). Critical water quality conditions such as depletion of dissolved oxygen are usually associated with low flow periods, and if this coincides with the summer season when the biological activity is most prominent, this can define critical conditions in terms of the maintenance of dissolved oxygen.

Expectancy of a particular drought flow is determined by plotting the drought flow data on log-extremal probability paper according to the method of Gumbel (IHD-WHO Working Group on the Quality of Water, 1978).   Figure 4 illustrates the procedure for plotting low flow data on such graph paper. Details of the construction of the log-extremal probability paper and other methods of plotting are cited in most reference works on river hydraulics.

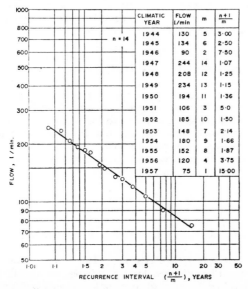

Figure 4:   Example of Drought Frequency Analysis (IHD-WHO Working Group on the Quality of Water, 1978)

Droughts with recurrent intervals of 5 to 10 years may be used as the basis for water quality management programs. While a complete analysis of drought flows may be beyond the scope of a particular river study, it does represent information which is necessary for the long-term management of the resource. Accordingly, effort should be made to either implement a program for making flow measurements and doing drought analysis, or this should be coordinated with other governmental agencies.

Mixing Analysis

Another important hydrological measurement relative to the location of sampling stations is the general pattern of stream mixing. When tributaries and effluent channels from industrial and municipal sources meet a stream, mixing with stream water may take place from within a fraction of a kilometer to several kilometers downstream of the confluence. Mixing will occur in three directions--vertical, lateral, and longitudinal. Since the flow in streams is almost invariably nonlaminar, vertical mixing is rapid and may take place within a few tenths of a kilometer. However, this may not be true in the cases of discharges of high salt-containing wastes or wastewaters with significantly different temperatures than that of the receiving stream. Thermal or density stratification may take place under these conditions and prevent rapid vertical mixing. However, due to the strong turbulence prevailing in most rivers, vertical uniformity can be assumed.

Lateral mixing is supported by the meandering of streams. Since most of the time wastewaters are discharged at the shoreline, the wastes stay close to one bank or the other. In swift streams lateral mixing may be further delayed. Lateral mixing is a function of the location of the outfall, direction of wastewater discharge, and velocity of the issuing wastewater with respect to the main current of the stream. In general, complete lateral mixing may be expected to take place within 1 to 3 kilometers downstream (IHD-WHO Working Group on the Quality of Water, 1978).

The variation of time-of-travel for individual water particles caused by different flow velocities over the river cross section, and simultaneous vertical and lateral movements, are responsible for longitudinal mixing. Time-variant pollution discharges into a river cause the most distinct variations in stream concentrations shortly after the discharge point. Further downstream these variations are smoothed out gradually due to longitudinal spreading of peaks of concentration. Longitudinal mixing does not influence location of sampling stations, however, its influence may be taken into account while interpreting data. For streams not influenced by tidal waters, the effect of longitudinal mixing in interpretation of data may be omitted.

WATER QUALITY PARAMETERS

A critical component of the study of contamination of a river is associated with the parameters which will be monitored for determining the baseline water quality characteristics. There are hundreds of water constituents which could be selected for monitoring, and it will be necessary to compromise between the number of parameters measured and the marginal difference that information on each parameter makes to subsequent interpretation and management decisions. As noted in Chapter 1, water quality can be described in terms of physical, chemical and bacteriological parameters.

One approach which can be used in the selection of water quality parameters to be monitored is to utilize general recommendations for conduction of water quality surveys. Table 9 contains a reference list of parameters used for river water quality surveys (IHD-WHO Working Group on the Quality of Water, 1978). Table 9 is structured into parameters proposed for inclusion in all surveys, those recommended for surveys with emphasis on the collection of baseline data, and those recommended for studies involving the monitoring of municipal and/or industrial pollution. In addition, optional parameters for other special surveys are also identified.

Another approach which can be used to identify potential water quality parameters for a study river is to identify the types of pollutants anticipated from waste discharges. For example, consider petroleum development and operation in the study area. This approach would be based on monitoring for those specific pollutants anticipated to be discharged into the river, either directly or indirectly, as a result of the developments planned for the area. Table 10 summarizes the concentrations of final effluent water pollutants found during a screening study in petroleum refining wastewater (Canter, 1981). Table 11 summarizes the concentrations of various toxic pollutants found in the final effluents from petroleum refining wastewater treatment facilities, with this information also indicative of the types of pollutants which could be discharged into the study river, and for which baseline information would be desirable (Canter, 1981).

Table 10:   Concentrations of Conventional Pollutants Found During a Screening Study in Petroleum Refining Wastewater (Canter, 1981)

Pollutant	Final Effluent		
	Number of Values	Range	Median
BOD_5[a], mg/l	43	<1 – 210	<10
COD, mg/l	48	28 – 820	120
TOC, mg/l	47	7 – 290	36
TSS, mg/l	47	2 – 110	21
Ammonia, mg/l	48	<1 – 53	5.0
Cr^{+6}, mg/l	45	0.01 – 0.11	<0.02
S^{-2}, mg/l	49	<0.1 – 2.1	0.5
Oil and grease, mg/l	27	3 – 53	13
pH	48	6.9 – 8.8	7.7

Table 9:   Selection of Parameters for River Water-Quality Surveys (IHD-WHO Working Group on the Quality of Water, 1978)

| Type of survey | Physical parameters | Chemical parameters | | | Biological parameters | |
		Inorganic	Organic	Nutrients	Microbiological	Hydrobiological
Proposed for inclusion in all surveys	Colour pH Specific conductance Suspended solids Total solids		Chemical Oxygen Demand (COD) Total Organic Carbon (TOC)		Coliforms, total and faecal	
Recommended for collection of baseline data	Odour	Acidity Alkalinity Calcium, Ca Chlorides, Cl Dissolved oxygen Hardness Iron, Fe Magnesium, Mg Manganese, Mn Potassium, K Selenium, Se Silver, Ag Sodium, Na	Biochemical Oxygen Demand (BOD); immediate, 5-day, ultimate	Nitrate nitrogen, NO_3	Total plate count	
Recommended additional parameters where municipal and/or industrial pollution are expected	Floating solids	Arsenic, As Barium, Ba Beryllium, Be Boron, B Cadmium, Cd Chromium, Cr Copper, Cu Dissolved Carbon Dioxide, CO_2 Fluorides, F Hydrogen sulphide, H_2S Lead, Pb Mercury, Hg Nickel, Ni Vanadium, V Zinc, Zn	Cyanide, CN Dissolved organic carbon Methylene Blue Active Substances (MBAS) Oil and grease Pesticides Phenolics	Ammonia nitrogen, NH_4 Nitrite nitrogen, NO_2 Organic nitrogen Soluble phosphorus Total phosphorus	Faecal streptococci Salmonella	Benthos Plankton counts
Optional parameters for surveys of special purpose	Bed load Light penetration Particle size Sediment concentration Settleable solids	Aluminium, Al Sulphates	Carbon Alcohol Extract (CAE) Carbon Chloroform Extract (CCE) Chlorine demand	Organic phosphorus Orthophosphates Polyphosphates Reactive silica	Shigella Viruses: —Coxsackie A&B —Polio —Adenoviruses —Echoviruses	Chlorophylls Fish Periphyton Taxonomic composition

Table 10: (continued)

Pollutant	Final Effluent		
	Number of Values	Range	Median
Cyanides, mg/l	53	<0.005 - 0.80	<0.03
Phenols, mg/l	49	<0.001 - 0.080	0.013
BOD$_5$[b], mg/l	38	<1 - 92	7
Flow, mgd	38	0.017 - 17.6	2.27

Note: Blanks indicate data not available.

[a]Seed from domestic sewage treatment plant.

[b]No seed.

Table 11:   Concentrations of Toxic Pollutants Found During a Screening
Study in Petroleum Refining Wastewater (Canter, 1981)

Toxic Pollutants	Final Effluent[a]		
	Number of Values[b]	Range	Median
Metals and Inorganics			
Antimony[c]	17	<1 - <25	<25
Arsenic[c]	18	<4 - 800	<20
Asbestos[d]	4		ND
Beryllium[c]	85	<1 - <3	<2
Cadmium[c]	86	<1 - 20	<20
Chromium[c]	87	1 - 1,200	50
Copper[c]	85	3 - 300	6
Cyanide	59	<5 - 320	<30
Lead[c]	87	2 - 210	<60
Mercury[c]	73	<0.1 - 12.0	<0.5
Nickel[c]	89	<1 - 74	20
Selenium[c]	31	3 - 32	19
Silver[c]	84	<1 - <25	<25
Thallium	32	<1 - <15	4
Zinc[c]	92	<10 - 1,300	84

Table 11: (continued)

Toxic Pollutants	Final Effluent[a]		
	Number of Values[b]	Range	Median
Pthalates			
Bis(2-ethylhexyl) phthalate	6	<10 - 2,000	450
Di-n-butyl phthalate	6	ND - 10	ND
Diethyl phthalate	4	ND - 30	1
Dimethyl phthalate	3	ND - 3	ND
Phenols			
2-Chlorophenol	1		ND
2,4-Dichlorophenol	1		10
2,4-Dinitrophenol	3		ND
2,4-Dimethylphenol	10	ND - <10	ND
2-Nitrophenol	1		ND
4-Nitrophenol	4	ND - <10	ND
Pentachlorophenol	1		ND
Phenol	15	ND - <10	ND
4,6-Dinitro-o-cresol	1		ND
Parachloromete cresol	3	ND - 10	<10
Aromatics			
Benzene	12*	ND - 12	<10
1,2-Dichlorobenzene	2		ND
1,4-Dichlorobenzene	2		ND
Ethylbenzene	9	ND - <10	ND
Toluene	13	ND - 35	ND
Polycrylic Aromatic Hydrocarbons			
Acenaphthene	7	ND - 6	ND
Acenaphthylene	5		ND
Anthracene	2		ND
Benzo(a)pyrene	2	1.3 - 3	2.2
Crysene	9	ND - 1.4	ND
Fluoranthene	9	ND - <0.1	ND
Fluorene	5		ND
Naphthalene	11	ND - 0.1	ND
Phenanthrene	12	ND - 1	ND
Pyrene	7	ND - 7	<0.1
Polychlorinated Biphenyls and Related Compounds			
Aroclor 1016	7	ND - <10	<10
Aroclor 1221	7	ND - <10	<10

Table 11: (continued)

| Toxic Pollutants | Final Effluent [a] | | |
	Number of Values[b]	Range	Median
Aroclor 1232	8	ND - <10	<5
Aroclor 1242	9	ND - <10	ND
Aroclor 1248	4		<10
Aroclor 1254	4		<10
Aroclor 1260	4		<10
Halogenated Aliphatics			
Carbon tetrachloride	4	ND - <10	<10
Chloroform	8[a]	ND - 66	ND
Dichlorobromomethane	1		ND
1,2-Dichloroethane	2	ND - <10	<5
1,2-Trans-dichloroethane	3	ND - <10	ND
Methylene chloride	8[e]	ND - >100	<70
1,1,2,2-Tetrachloroethane	2		<10
Tetrachloroethylene	4	ND - <10	<5
1,1,1-Trichloroethane	1		ND
Trichloroethylene	2	ND - <10	<5
Pesticides and Metabolites			
Aldrin	2		ND
α-BHC	1		ND
β-BHC	3		ND
δ-BHC	2		ND
γ-BHC	1		ND
Chlordane	1		ND
4,4'-DDE	1		ND
4,4'-DDD	1		ND
α-Endosulfan	1		ND
β-Endosulfan	1		ND
Endosulfan sulfate	1		ND
Heptachlor	2	ND - <5	<2.5
Heptachlor epoxide	2		ND
Isophorone	2		ND

Note: Blanks mean compound data not available or, if in "range" column, range is not defined.   ND—Not detected in sample.

[a]Values are corrected for blanks when blank values are reported.

Table 11: (continued)

---

bValues include samples which contained nondetectable quantities.

cValues include 3-day composite samples.

dUnits of million fibers per liter.

eNot all values counted because values in blanks greater than values in sample(s).

In order to analyze the planned water quality parameters to be monitored in a river survey relative to the general parameters listed in Tables 9 through 11, several additional tables will be presented. Table 12 summarizes the physical parameters identified in Tables 9 and 10. Table 13 summarizes the inorganic parameters included in Tables 9 through 11. Specific trace inorganics in the effluents from petroleum refineries include arsenic, asbestos, beryllium, copper, selenium, silver, and thallium. These constituents are generally present in trace concentrations only, and unless there would be other industrial sources, an intensive monitoring program for these specific elements would probably not be necessary. However, it might be desirable to conduct periodic tests for these specific parameters at selected sampling stations at a later time in the survey.

Table 12: Physical Water Quality Parameters

Identified Parameters

General Surveys (Table 9)	Petroleum Industry Effluent (Table 10)
Color	
pH	pH
Specific Conductance	
Suspended Solids	Suspended Solids
Total Solids	
Odor	
Floating Solids	

Table 13: Inorganic Water Quality Parameters

Identified Parameters

General Surveys (Table 9)	Petroleum Industry Effluent (Table 10 and 11)

Acidity	
Alkalinity	
	Antimony
Arsenic	Arsenic
	Asbestos
Barium	
Beryllium	Beryllium
Boron	
Cadmium	Cadmium
Calcium	
Chlorides	
Chromium	Chromium
Copper	Copper
Dissolved Carbon Dioxide	
Dissolved Oxygen	
Fluorides	
Hardness	
Hydrogen Sulphide	
Iron	
Lead	Lead
Magnesium	
Manganese	
Mercury	Mercury
Nickel	Nickel
Potassium	
Selenium	Selenium
Silver	Silver
Sodium	
	Sulfites
	Thallium
Vanadium	
Zinc	Zinc

The inorganic nutrients are listed in Table 14 as a summary of the nutrients contained in both Tables 9 and 10. Consideration should be given to occasional monitoring for organic nitrogen, even though this would only be suspected primarily in areas downstream from existing sources of municipal or industrial pollution. Organic water quality constituents are summarized in Table 15. There are numerous trace organics which might be found in petroleum refinery effluents, with a listing of these found in Table 16 (Canter, 1981). It is not suggested that a major monitoring effort be undertaken for these specific organics; however, this information should be considered in planning longer-term studies of river quality.

Table 14: Nutrient Water Quality Parameters

Identified Parameters	
General Surveys (Table 9)	Petroleum Industry Effluent (Table 10)
Nitrates	
Ammonia	Ammonia
Nitrites	
Organic Nitrogen	
Soluble Phosphorus	
Total Phosphorus	

Table 15: Organic Water Quality Parameters

Identified Parameters	
General Surveys (Table 9)	Petroleum Industry Effluent (Table 10 and 11)
Chemical Oxygen Demand	COD
Total Organic Carbon	TOC
Biochemical Oxygen Demand	
Immediate	
5-day	BOD_5
Ultimate	
Cyanides	Cyanides
Dissolved Organic Carbon	
Methylene Blue	
Active Substances	

Table 15: (continued)

Identified Parameters

General Surveys (Table 9)	Petroleum Industry Effluent (Table 10 and 11)
Oil and Grease	Oil and Grease
Pesticides	
Phenolics	Phenols

Table 16:   Additional Organics Which May Be Found in Petroleum Refinery Effluents (Canter, 1981)

Pthalates

    Bis(2-ethylhexyl) phthalate
    Di-n-butyl phthalate
    Diethyl phthalate
    Dimethyl phthalate

Phenols

    2-Chlorophenol
    2,4-Dichlorophenol
    2,4-Dinitrophenol
    2,4-Dimethylphenol
    2-Nitrophenol
    4-Nitrophenol
    Pentachlorophenol
    Phenol
    4,6-Dinitro-o-cresol
    Parachlorometa cresol

Aromatics

    Benzene
    1,2-Dichlorobenzene
    1,4-Dichlorobenzene
    Ethylbenzene
    Toluene

Polyacrylic aromatic hydrocarbons

    Acenaphthene
    Acenaphthylene

Table 16: (continued)

Anthracene
Benzo(a)pyrene
Chrysene
Fluoranthene
Fluorene
Naphthalene
Phenanthrene
Pyrene

Polychlorinated biphenyls and related compounds

Aroclor 1016
Aroclor 1221
Aroclor 1232
Aroclor 1242
Aroclor 1248
Aroclor 1254
Aroclor 1260

Halogenated aliphatics

Carbon tetrachloride
Chloroform
Dichlorobromomethane
1,2-Dichloroethane
1,2-Trans-dichloroethylene
Methylene chloride
1,1,2,2-Tetrachloroethane
Tetrachloroethylene
1,1,1-Trichloroethane
Trichloroethylene

Pesticides and metabolites

Aldrin
α-BHC
β-BHC
δ-BHC
γ-BHC
Chlordane
4,4'-DDE
4,4'-DDD
α-Endosulfan
β-Endosulfan
Endosulfan sulfate
Heptachlor
Heptachlor epoxide
Isophorone

One of the issues of concern in selecting water quality parameters is related to the analytical requirements for measurement. Table 17 contains a list of references related to analytical techniques for water quality parameters, with their abstracts found in Appendix E. Standard approaches for measuring parameters should be utilized (American Water Works Association and Water Pollution Control Federation, 1981).

Table 17:   References on Analytical Techniques for Water Quality
             Parameters

Authors (Year)	Comments
Baltisberger (1975)	Microdetermination of cationic forms of mercury.
Baltisberger (1977)	Differentiation of inorganic and organic mercury cations.
Bishop et al. (1978)	Anion exchange method for plutonium in water.
Chakravorty and Van Gricken (1982)	Determination of trace metals in water via co-precipitation with iron hydroxide and x-ray fluorescence analysis.
Chatfield, Dillon and Stott (1983)	Sample preparation techniques for asbestos fiber concentrations in water.
Cook, Duvall and Bourke (1978)	Improved methods for oil-in-water analyses.
Crowther (April 1978)	Autoclave digestion procedure for determination of total iron concentrations.
Crowther (July 1978)	Autoclave digestion and colorimetric procedure for total manganese concentrations.
Downes (1978)	Hydrazine reduction method for automated determination of low nitrate levels.
Elsenreich and Hullet (1979)	Isolation and quantitation of free and bound fatty acids in river water.
Favretto, Stancher and Tunis (1983)	Spectrophotometric determination of polyoxyethylene nonionic surfactants.
Johnston and Herron (1979)	Recovery of nonvolatile mutagenic compounds from surface waters.
Kloosterboer and Goossen (1978)	Total phosphate determination following photochemical decomposition and acid hydrolysis.

Table 17: (continued)

Authors (Year)	Comments
Lantz, Davenport and Wynveen (1980)	Automated electrochemical analysis for total organic carbon and chemical oxygen demand.
Manahan et al. (1973)	Analyses for low levels of complexing agents, including copper and cyanide.
Manning (1979)	Use of atomic absorption spectroscopy for inorganics.
Marks (1982)	Ion selective electrodes for water analyses.
Matsumoto (1981)	Fatty acids in polluted and unpolluted waters.
National Environmental Research Center (1971)	Handbook of chemical analytical procedures for water and wastewater.
Rock and Post (1980)	Field analysis of trichloroethylene in water.
Salim and Cooksey (1981)	Use of Coulter Counter for determining particle size distributions in water.
Seitz (1973)	Flame spectrometer determination of phosphorus in water.
Sekerka and Lechner (1982)	Determination of organohalides in water.
Sharma, Rathore and Ahmed (1983)	Colorimetric spot test for malathion residues in water.
Slabbert and Morgan (1982)	Bioassay technique for rapid assessment of toxicants in water.
Stepanenko and Muslova (1978)	Chromatographic determination of organic carbon in water.
Thurnau (1978)	Ion selective electrodes for water analyses.
Ullman (1976)	Use of laser Raman spectroscopy in water quality analyses.
Water Research Center (1977)	Simultaneous analysis of trace metals in water.

Microbiological parameters associated with general water quality monitoring are summarized in Table 9. Total coliforms represent a general indicator; however, consideration should be given to the measurement of fecal coliforms if evidences of existing municipal or industrial pollution are found in the study river. The analytical requirements for various microbiological parameters are also important, and Table 18 contains a pertinent list of references. The abstracts for the references are in Appendix F. Standard methods and procedures should be utilized (American Water Works Association and Water Pollution Control Federation, 1981).

Table 18: References on Bacteriological Water Quality Monitoring

Authors (Year)	Comments
Berman, Rohr and Safferman (1980)	Recirculating flow molecular filtration system for concentrating poliovirus 1.
Block and Rolland (1979)	Method for concentrating Salmonella from water.
Chappelle et al. (1983)	Bioluminescent assay for ATP in water-borne bacteria.
Fleischer and McFadden (1980)	Experimental design for improved coliform enumeration via the membrane filtration technique.
Hirn and Pekkanan (1977)	Bacterial preservation with various media.
Jeffers (1978)	Continuous measurement of the bacterial content of water samples.
Joshi et al. (1978)	Use of indigenously available peptones as media in membrane filter technique.
Lesar and Standridge (1979)	Sampling and storage methodology for sulfate reducing bacteria.
Manja, Maurya and Rao (1982)	Field test for screening of drinking water for fecal pollution.
Martins, Alves and Sanchez (1982)	Comparative study of Pseudomonas aeruginosa enumeration methods.
Reasoner, Blannon and Geldreich (1979)	Rapid test for detection of fecal coliforms in water.
Schillinger, Evans and Stuart (1978)	Limulus lysate assay for rapid determination of bacteriological water quality.

BIOLOGICAL PARAMETERS

Selection of aquatic biological parameters is complicated due to system relationships in terms of material and energy flows. Figure 5 is a schematic diagram indicating material and energy flows in an aquatic ecosystem along with system inputs and outputs (Canter, 1979). Aquatic organisms include primary producers, plant eaters, meat eaters, and decomposers. Primary producers include algae, while plant eaters encompass zooplankton, fish, and benthic organisms. Meat eaters include zooplankton and fish as well as benthic organisms. The monitoring program for a study river should include consideration of sampling of various planktonic and benthic forms, as well as fish.

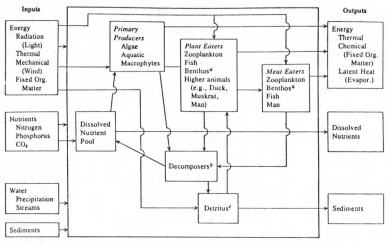

a. Organisms living at or on the bottom of bodies of water.
b. Fungi and bacteria.
c. Small particles of organic matter.

Figure 5: Material and Energy Flows in an Aquatic Ecosystem (Canter, 1979)

A nonpolluted river would be characterized by certain types and abundances of aquatic organisms based on climatological information as well as stream hydrological characteristics. When man-made pollution occurs there are changes in both the diversity and abundance of aquatic organisms. To illustrate pollutional effects upon aquatic organisms, Figure 6 displays the changes in the kinds and populations of animals as a result of organic pollution, toxic pollution, and silt (Nemerow, 1974). Numerous stream pollution studies have been conducted within the last several decades, and a general pattern of biological responses to pollution can be delineated.

Response of Organisms to Organic Pollution

To serve as an illustration of the changes in aquatic organisms resulting from organic pollution, untreated domestic sewage can be used as an example pollutant. With raw sewage, the lowering of dissolved oxygen and formation of sludge deposits are the most commonly occurring of the environmental

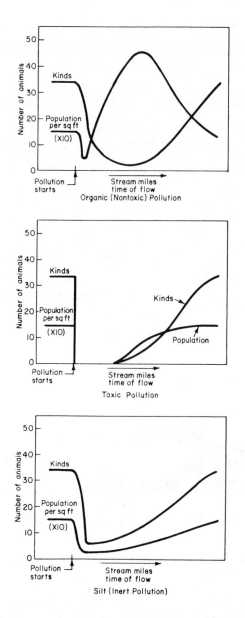

Figure 6: Pollutional Effects on Aquatic Animals (Nemerow, 1974)

alterations that damage aquatic biota (Bartsch and Ingram, 1959).   No two rivers are ever alike in detail, and hence the exact extrapolation of information on biological changes from one survey to usage on another river survey would be impossible.   However, general principles can be identified based on a hypothetical stream which is made to conform exactly to theory, with the response of the stream and its biota representing typical responses under ideal conditions.   Application of these general principles can then be made to the biological sampling portion of the study river.   The following information is from a classic stream pollution study widely used in environmental engineering for over two decades (Bartsch and Ingram, 1959).

Figure 7 demonstrates the oxygen characteristics along with biochemical oxygen demand in a hypothetical stream (Bartsch and Ingram, 1959).   In this situation raw domestic sewage from a sewered community of 40,000 people flows into the stream.   The volume flow in the stream is 100 cubic feet per second, complete mixing is assumed, and the water temperature is 25⁰ Centigrade.   Under these conditions the dissolved oxygen sag curve reaches a low point after 2¼ days of flow, and then rises again toward a restoration similar to that of the upstream, unpolluted water.   The biochemical oxygen demand is low upstream, increases immediately downstream from the pollutant discharge, and then follows a typical pattern of reduction as a result of biological decomposition within the stream.   From left to right on Figure 7 the stream zones can be defined as clean water, degradation, active decomposition, recovery, and clean water.   As bacteria decompose organic matter in the degradation and active decomposition zones a demand is placed on the dissolved oxygen resources within the stream.   If there were no other resources to replenish the dissolved oxygen, the dissolved oxygen in the stream would drop to 0 within approximately 18 miles downstream.   However, natural reaeration occurs and the oxygen within the stream is replenished as a result of transfer from the atmosphere.   Accordingly, there is a recovery zone so that the dissolved oxygen will again return to its upstream levels.

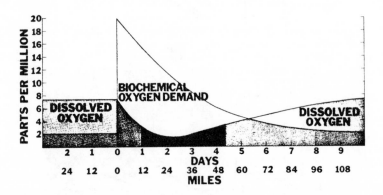

Figure 7:   Effects of Organic Pollution on Dissolved Oxygen (Bartsch and Ingram, 1959)

Figure 8 illustrates the effects of organic pollution on aquatic plants (Bartsch and Ingram, 1959).   Sludge deposits begin to accumulate just below the point of sewage discharge.   These deposits reach their maximum thickness near the point of origin, but blanket the stream bed for many miles downstream. The substance of the deposits gradually is reduced by decomposition through the action of bacteria, moulds, and other sludge-dwelling organisms, until it becomes insignificant about 30 miles below the municipality.   Also, at the outfall the water is turbid from fine solids held in suspension by the flowing water.   Larger floating solids, destined to sink eventually to the stream bed as settlable solids, are visible on the water surface as they drift downstream.

Both the fine and large solids contribute to the sludge deposit, and as they settle progressively to the bottom of the stream bed, the water becomes clear and approaches the color and transparency of upstream water above the point of sewage discharge.

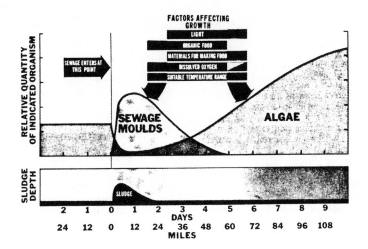

Figure 8:   Effects of Organic Pollution on Aquatic Plants (Bartsch and Ingram, 1959)

The upper portion of Figure 8 illustrates the relative distribution and quantities of algae and various moulds. From mile 0 to mile 36, high turbidity from floating debris and suspended solids is not conducive to algae production. Thus, except for slimy blue-green marginal and bottom types, algae are sparse in this reach. In order to grow well algae need sunlight, and here it cannot penetrate the water effectively. Also, floating solids that settle out of the water carry to the bottom with them floating algae that drift into the polluted zone from clear water areas upstream. The forms of algae associated with clean water, as well as those common in organically enriched areas are summarized in Table 19 (Nemerow, 1974). Classifications of algae from the study river could be compared to the information in Table 19 for an indication as to whether or not the species were indicative of clean water or organically enriched areas.

Figure 9 illustrates the interrelations between bacteria and animal plankton, such as ciliated protozoans, rotifers, and crustaceans (Bartsch and Ingram, 1959). The quantities shown and the die-off curves for sewage bacteria and for coliform bacteria separately are theoretically accurate. The center curve for ciliated protozoans and the last curve representing rotifers and crustaceans are more accurate in principle than in actual quantities.

Figure 10 illustrates the types of organisms and the numbers of each type likely to occur along the course of the stream under the assumed physical conditions stated earlier (Bartsch and Ingram, 1959). The upper curve represents the numbers of species of higher form organisms found under varying degrees of pollution. The lower curve represents the number of individuals of a species. In clean water above the city a great variety of organisms is found

Table 19: Classification of Clean Water and Pollution Algae (Nemerow, 1974)

Clean Water Algae	Pollution Algae-Algae Common in Organically Enriched Areas
Group and Algae	Group and Algae

Blue-Green Algae *(Myxophyceae):* *Agmenellum quadriduplicatum, glauca type* *Calothrix parietina* *Coccochloris stagnina* *Entophysalis lemaniae* *Microcoleus subtorulosus* *Phormidium inundatum* Green Algae *(Nonmotile Chlorophyceae):* *Ankistrodesmus falcatus, var. acicularis* *Bulbochaete mirabilis* *Chaetopeltis megalocystis* *Cladophora glomerata* *Draparnaldia plumosa* *Euastrum oblongum* *Gloeococcus schroeteri* *Micrasterias truncata* *Rhizoclonium hieroglyphicum* *Staurastrum punctulatum* *Ulothrix aequalis* *Vaucheria geminata* Red Algae *(Rhodophyceae):* *Batrachospermun vagum* *Hildenbrandia rivularis* *Lemanea annulata* Diatoms *(Bacillariophyceae):* *Amphora ovalis* *Cocconeis placentula* *Cyclotella bodanica* *Cymbella cesati* *Meridion circulare* *Navicula exigua var. capitata* *Navicula gracilis* *Nitzschia linearis* *Pinnularia nobilis* *Pinnularia subcapitata* *Surirella splendida* *Synedra acus var. angustissima* Flagellates *(Chrysophyceae, Cryptophyceae, Euglenophyceae and Volvocales of Chlorophyceae):* *Chromulina rosanoffi* *Chroomonas nordstetii* *Chroomonas setoniensis* *Chrysococcus major*	Blue-Green Algae *(Myxophyceae):* *Agmenellum quadriduplicatum, tenuissima type* *Anabaena constricta* *Anacystis montana* *Arthrospira jenneri* *Lyngbya digueti* *Oscillatoria chalybea* *Oscillatoria chlorina* *Oscillatoria formosa* *Oscillatoria lauterbornii* *Oscillatoria limosa* *Oscillatoria princeps* *Oscillatoria putrida* *Oscillatoria tenuis* *Phormidium autumnale* *Phormidium uncinatum* Green Algae *(nonmotile Chlorophyceae):* *Chlorella pyrenoidosa* *Chlorella vulgaris* *Chlorococcum numicola* *Scenedesmus quadriccula* *Spirogyra communis* *Stichococcus bacillaris* *Stigeoclonium tenue* *Tetraedron muticum* Diatoms *(Bacillariophyceae):* *Gomphonema parvulum* *Hantzschia amphioxys* *Melosire varians* *Navicula cryptocephala* *Nitzschia acicularis* *Nitzschia palea* *Surirella ovata* Flagellates *(Euglenophyceae, Volvocales of Chlorophyceae):* *Carteria multifilis* *Chlamydomonas reinhardi* *Chlorogonium euchlorum* *Cryptoglena pigra* *Euglena agilis* *Euglena deses* *Euglena gracilis* *Euglena oxyuris* *Euglena polymorpha* *Euglena viridis* *Lepocinclis ovum*

Table 19: (continued)

Clean Water Algae	Pollution Algae-Algae Common in Organically Enriched Areas
Group and Algae	Group and Algae
Chrysococcus ovalis *Chrysococcus rufescens* *Dinobryon stipitatum* *Euglena ehrenbergii* *Euglena spirogyra* *Mallomonas caudata* *Phacotus lenticularis* *Phacus longicauda* *Rhodomonas lacustris*	*Lepocinclis texta* *Pandorina morum* *Phacus pyrum* *Pyrobotrys gracilis* *Pyrobotrys stellata* *Spondylomorum quaternarium*

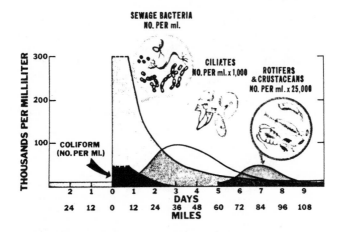

Figure 9:   Effects of Organic Pollution on Selected Aquatic Biota (Bartsch and Ingram, 1959)

with very few of each kind represented. At the point of waste entry the number of different species is greatly reduced, and they are replaced by different associations of aquatic life. This new association demonstrates a severe change in environment that is drastically illustrated by a change in the species make-up of the biota. However, this changed biota, represented by a few species, is accompanied by a tremendous increase in the numbers of individuals of each kind as compared with the density of population upstream.

In the clean zone water upstream there is an association of sports fish, various minnows, caddis worms, mayflies, stoneflies, hellgrammites, and gill-breathing snails, with each kind represented by a few individuals. In badly polluted zones the upstream association disappears completely, or is reduced, and is replaced by a dominant animal association of rat-tailed maggots, sludge

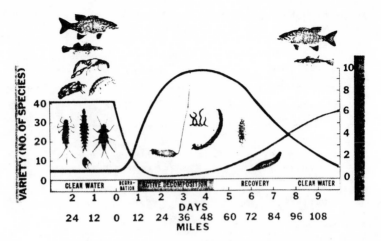

Figure 10:   Effects of Organic Pollution on Higher Aquatic Fauna (Bartsch and Ingram, 1959)

worms, blood worms, and a few others, with each represented by a great number of individuals. When downstream conditions again resemble those of the upstream clean water zone, the clean water animal association tends to reappear and the pollution tolerant group of animals become suppressed. Table 20 provides a comparison of clean water animal forms with those commonly associated with polluted waters (Nemerow, 1974).

Table 20: Organism Associations in Clean and Polluted Waters (Nemerow, 1974)

	Clean water association		Polluted water association
Algae	*Cladophora* (green)	Iron bacteria . .	*Sphaerotilus*
	Ulothrix (green)	Fungi	*Leptomitus*
	Navicula (diatom)	Algae	*Chlorella* (green)
Protozoa	*Trachelomonas*		*Chlamydomonas* (green)
Insects	Plecoptera (stoneflies in		*Oscillatoria* (blue-green)
	general)		*Phormidium* (blue-green)
	Negaloptera (hellgram-		*Stigeoclonium* (green)
	mites, alderflies, and		
	fishflies in general	Protozoa	*Carchesium* (stalked
	Trichoptera (caddisflies		colonial ciliate)
	in general)		*Colpidium* (non-colonial
	Ephemeroptera (may-		ciliate)
	flies in general)	Segmented . .	*Tubifex* (sludgeworms)
	Eimidae (riffle beetles	Worms	*Limnodrilus* (sludgeworms)
	in general)	Leeches	*Helobdella stagnalis*
Clams	Unionidae (pearl button)	Insects	*Culex pipiens* (mosquito)
Fish*	*Etheostoma* (darter)		*Chironomus* (-Tendpipes)
	Notropis (shiner)		*plumosus* (bloodworms)
	Chrosomus (dace)		*Tubifera (Eristalis tenax)*
			(rat-tailed maggot)
		Snail	*Physa integra*
		Clam	*Sphaerium* (fingernail)
		Fish*	*Cyprinus carpio* (carp)

* Names from: American Fisheries Society Special Publication No. 2, "A List of Common and Scientific Names of Fishes from the United States and Canada" (Second Edition) Ann Arbor, Mich. (1960), 102 pp.

Figure 11 contains an additional diagrammatic view of the succession of organisms resulting from organic matter introduced in a stream. There are a series of changes in organism types as the degradation process proceeds (IHD-WHO Working Group on the Quality of Water, 1978).

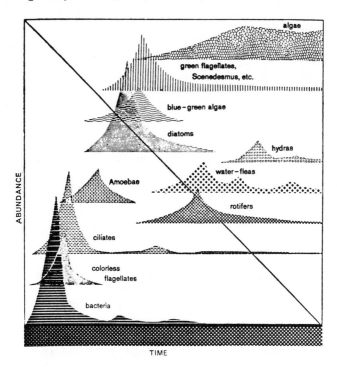

Figure 11:   Succession of Organisms in the Degradation of Organic Matter (IHD-WHO Working Group on the Quality of Water, 1979)

Classification of Streams

Several systems for classifying streams based on their biological characteristics have been developed. Three examples will be presented, with the first focused on a basic delineation of four stream segments (Ballinger, 1963; Jackson, 1963a), the second based on five classifications (Nemerow, 1974), and the third based on three zones (Nemerow, 1974).

Based on the work by Bartsch and Ingram (1959), four major stream zones can be identified (Ballinger, 1963; Jackson, 1963a). These four zones are the unpolluted stream, zone of recent pollution, septic zone, and recovery zone. The unpolluted or clean water zone would represent an area that had no evidence of organic pollution. The dissolved oxygen concentrations would be high and the biochemical oxygen demand would be low. The organisms are present in a wide variety, fish are reasonably abundant, and the general visual appearance of the stream is pleasing. Table 21 summarizes the general features and characteristic fauna of the clean water zone (Jackson, 1963a).

Table 21: Characteristics of Stream Zones (Jackson, 1963a)

Zone and General Features	Characteristic Fauna
Clean Water Zone	
a. DO high	a. Caddis fly larvae (Trichoptera)
b. BOD low	b. Mayfly larvae (Ephemeroptera)
c. Turbidity low	c. Stonefly larvae (Plecoptera)
d. Organic content low	d. Damselfly larvae (Zygoptera)
e. Bacterial count low	e. Beetles (Coleoptera)
f. Numbers of species high	f. Clams (Pelecypoda)
g. Numbers of organisms of each species moderate or low.	g. Fish such as: Minnows, Notropid type; Darters (Etheostomidae); Millers thumbs (Cottidae); many sunfishes and basses (Centrarchidae); Sauger, yellow perch, etc. (Percidae); others
h. Bottom free of sludge deposits	
Degeneration or Recent Pollution	
a. DO variable, 2 ppm to saturation	a. Midge larvae (Chironomidae or Tendipedidae) becoming more abundant.
b. BOD high	
c. Turbidity high	b. Back swimmers (Corixidae) and water boatmen (Notonectidae) often present
d. Organic content high	
e. Bacterial count variable to high	c. Sludge worms (Tubificidae) common
f. Number of species declining from clean water zone	d. Dragonflies (Anisoptera) often present have unique tail breathing strainer
g. Number of organisms per species tends to increase	e. Fish types: Fathead minnows (Pimephales promelas); White sucker (Catostomus commerson- nii); Bowfin (Amia calva); Carp (Cyprinus carpio)
h. Other: Slime appearing on bottom	

Table 21: (continued)

Zone and General Features	Characteristic Fauna

Septic or Putrefaction

a. DO little or none during warm weather

b. BOD high but decreasing

c. Turbidity high, dark, odoriferous

d. Organic content high but decreasing

e. Bacterial count high

f. Number of species very few

g. Number of organisms: may be extremely high or none at all

h. Other: Slime blanket and sludge deposits usually present, oily appearance on surface, rising gas bubbles

a. Mosquito larvae

b. Rat-tailed maggots

c. Sludge worms (Tubificidae)

d. Air breathing snails (Physa)

e. Fish types: None

Recovery

a. DO 2 ppm to saturation

b. BOD dropping

c. Turbidity dropping, less color and odor

d. Organic content dropping

e. Bacterial count dropping

f. Numbers of species increasing

g. Numbers of organisms per species decreasing, (with the increase in competition)

h. Other: less slime and sludge

a. Midge larvae (Chironomids)

b. Black fly larvae (Simulium)

c. Giant Water bugs (Belostoma spp.)

d. Clams (Unio)

e. Fish types: Green sunfish (Lepomus cyanellus); Common sucker (Catostomus commersonnii); Flathead catfisher (Pilodictis olivaris); Stoneroller minnow (Campostoma anomalum); Buffalo (Megastomatobus cyprinella)

The zone of recent pollution, or the zone of degeneration, represents that area immediately downstream from the point of wastewater discharge. In this area the stream dissolved oxygen begins to decrease and the biochemical oxygen demand is high. Sewage bacteria will predominate, although a few other types of organisms may be present. The stream is generally discolored and may contain quantities of floating solids. General features and characteristic fauna of the zone of recent pollution are listed in Table 21 (Jackson, 1963a).

The septic zone, or zone of putrefaction, is further downstream from the point of wastewater discharge, and active decomposition is characteristic of this area. Dissolved oxygen is at a minimum and there may be associated odors due to hydrogen sulfide. Few species of organisms are present but they exist in large numbers. This zone is the most aesthetically displeasing of the four zones. Table 21 contains a summary of the general features and characteristic fauna of the zone of putrefaction (Jackson, 1963a).

The recovery zone is further downstream from the septic zone and represents the area where evidences of natural stream purification are predominant. The dissolved oxygen in the stream is increasing and various organisms, including fish, will tend to be abundant. There may be problems in the recovery zone due to excessive algae growth resulting from the high nutrient concentrations. Table 21 also contains a summary of the general features and characteristic fauna of the zone of recovery (Jackson, 1963a). Further downstream from the zone of recovery, unless there are other wastewater discharges, the stream characteristics in terms of both water quality as well as characteristic fauna should be similar to those in the upstream or clean water zone.

Patrick developed a system of observing organisms in streams and assuming the degree of pollution from an analysis of the groups and relative numbers present (Nemerow, 1974). Patrick suggested seven different taxonomic groups of organisms to be used as biological measures of stream conditions, with these groups as follows: (1) the blue green algae, some green algae, and some rotifers; (2) oligochaetes, leeches, and snails; (3) protozoa; (4) diatoms, red algae, and most green algae; (5) all rotifers not in (1) above plus clams, worms, and some snails; (6) all insects and crustacea; and (7) all fish. From observations and enumerations of the seven groups, Patrick arrived at five classifications of a river as follows:

Healthy Stream--balance of organisms; the algae are mainly diatoms and green algae; the insects and fish are represented by a variety of species. Groups of organisms numbers (4), (6), and (7) above are all above the 50 percent level based on levels found in background stations upstream.

Semi-healthy Stream--the ecological balance is somewhat disrupted; the general patterns of water quality as well as aquatic organisms are irregular; a given species will be represented by a greater number of individuals. The following possibilities may appear--either groups (6) and (7) above will be below 50 percent of their level in a healthy stream, and (1) and (2) above will be below 100 percent of their levels in a healthy stream; or either (6) or (7) above will be below 50 percent and (1), (2), and (4) above will be at or above the 100 percent level for a healthy stream.

Polluted--this is a stream in which the balance of organisms, based on a healthy stream segment, is found to be disrupted. Conditions may be favorable for some groups of organisms such as (1) and (2) above. The following may also be observed--species of either or both (6) and (7) above are absent, and species (1) above is at 50 percent or greater in terms of a healthy stream. Species of (6) and (7) above may both be present but below 50 percent, and species (1) and (2) may be at 100 percent or more of their levels for a healthy stream.

Very Polluted--this is a stream segment in a condition that is definately toxic to plant and animal life. Many of the above groups of organisms may be entirely absent. This state will occur if groups (6) and (7) above are absent and (4) is below 50 percent of the healthy stream level; or if (6) and (7) are present but groups (1) or (2) are less than 50 percent of the healthy stream level.

Atypical--this particular classification cannot be compared to any of the above four stream groupings due to general ecological conditions or other unusual factors.   This may occur in areas that have eutrophic or oligotrophic conditions. Basically, this group is one that does not fit the normal pattern of stream pollution and recovery.

It should be noted that from the above list of seven taxonomic groups of organisms, groups (3) and (5) were not specifically used by Patrick in the five stream classifications. This would suggest that perhaps these two groups could be eliminated in terms of sampling programs (Nemerow, 1974).

Another system for classifying streams is related to the influence of the organic matter on biological species. This system can be referred to as the saprobicity system (Nemerow, 1974). Three specific zones can be defined based on the saprobicity system, with the term saprobicity referring to a measure of biodegradable organic matter. The specific zones are as follows:

Zone 1 Polysaprobic--the zone of gross pollution with high molecular weight organic matter; very little or no dissolved oxygen; formation of sulfides; abundant bacteria and other organisms; few species of animals living on decaying organic matter or bacteria.

Zone 2 Mesosaprobic--this zone contains simpler organic compounds; steadily increasing oxygen; the upper portion contains many bacteria and fungi with more types of animals, but few algae; the lower portion possesses more mineralization (the conversion of organic to inorganic matter) suitable for algae-intolerant animals and some rooted plants.

Zone 3 Oligosaprobic--the zone of recovery where mineralization is complete and oxygen is back to normal. It contains a wide range of plants and animals.

In summary regarding these classification systems, they all embody the concepts of changes in species diversity and abundance as a result of stream pollution.   They also recognize the fact that there are various types of organisms which can be monitored for use in surveys to establish the degree of contamination in a given system. Accordingly, this type of information should be utilized in the planning of a river biological monitoring program.

Biological Monitoring

Based on the general principles described above, the focus of a biological monitoring program should be on collecting several different species, characterizing them by types, and the gathering of information relative to their abundance. As a rule, ecological systems which offer a great variety of conditions are characterized by a great diversity of species, each present as relatively few individuals. Alternatively, a scarcity of species with an abundance of individuals is likely to be the case in systems where the range of living conditions is restricted or where these restrictions are brought about by pollutional conditions (IHD-WHO Working Group on the Quality of Water, 1978). A biological monitoring program for a river should probably contain emphasis on three groups of organisms--plankton, benthic organisms, and fish. Table 22 contains a list of references related to biological monitoring, with their respective abstracts found in Appendix G.

Table 22: References on Biological Monitoring

Authors (Year)	Comments
Bingham et al. (1982)	Grab samplers for benthic macroinvertebrates.
Cairns, Dickson and Westlake (1976)	Series of papers on biological monitoring of water and effluent quality.
Cairns and Gruber (1980)	Biological monitoring systems based on measuring the ventilatory behavior of fish.
Hellawell (1978)	Handbook for the biological monitoring of rivers.
Jacobs and Grant (1978)	Methods of zooplankton sampling and analysis for quantitative surveys.
Kingsbury and Rees (1978)	Rapid biological methods for continuous water quality monitoring.
National Technical Information Service (1983)	Use of molluscs to monitor water quality.
Roline and Miyahara (1979)	Use of algal assay bottle test in water quality studies.
Slooff (1977)	Fish respiration for biological monitoring of toxicants.
States et al. (1978)	Planning for ecological baseline studies related to energy development projects in the western United States.

Table 22: (continued)

Authors (Year)	Comments
Stout et al. (1978)	Planning for integrated baseline studies of the environment.
Ward (1978)	Book on the planning, conduction, and interpretation of biological impact studies.
Weber (1973)	Manual of biological field and laboratory methods.
Weiss (1976)	Evaluation of the algal assay bottle test.

Plankton refer to the microscopic plants and animals normally swimming or suspended in open water. Clean waters in general tend to be characterized by the presence of planktonic forms such as diatoms and certain types of green algae. Polluted waters are characterized by planktonic forms such as blue-green algae, protozoa, and certain green algae (Jackson, 1963a). Plankton sampling may involve the use of bottles, depth samplers, or plankton nets. Plankton nets concentrate the sample in the act of collecting and capture certain larger forms which escape from either bottles or depth samplers. Additional information on plankton sampling is in Chapter 6.

Benthic organisms, or bottom-dwelling fauna, are animals that live directly in association with the bottom of a stream. They may crawl on, burrow in, or attach themselves to the bottom. Macro-organisms are usually defined as those organisms that will be retained by a No. 30 sieve (a No. 30 U.S. standard sieve has openings of 0.59 mm and is formed from wire 0.29 to 0.42 mm in diameter). In essence, the organisms retained by the sieve are those visible to the unaided eye. Representative bottom-dwelling macrofauna are shown in Figure 12 (Keup, Ingram and Mackenthun, 1966).

Bottom-dwelling organisms have inherent qualities that make their use in pollution surveys advantageous. These qualities include a pronounced response to pollution, a sufficiently long life cycle to prevent a response to intermittent relief from pollution, and either a means of locomotion that prohibits extended rapid migrations or a sessile-attached mode of life that reduces the influence of neighboring water conditions on the organisms. Because of these qualities, bottom-dwelling organisms reflect conditions at the sampling point for an extended period of time. A wide variety of bottom-dwelling macro-organisms inhabit nonpolluted waters. Each occupies a niche in the benthic community, where the adaptations of these species are most efficiently utilized in maintaining life processes. Each is limited in numbers by availability of food supply, intra- and inter-species competition, predation, and the stage of its life cycle. Since all of these factors are affected by pollution, a biological survey of the bottom-dwelling macrofauna is in fact an investigation into the extent and degree of water pollution (Keup, Ingram and Mackenthun, 1966).

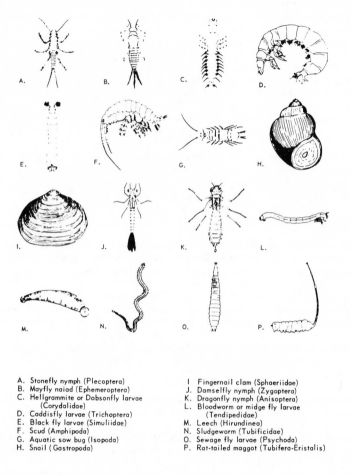

A. Stonefly nymph (Plecoptera)
B. Mayfly naiad (Ephemeroptera)
C. Hellgrammite or Dobsonfly larvae
   (Corydalidae)
D. Caddisfly larvae (Trichoptera)
E. Black fly larvae (Simuliidae)
F. Scud (Amphipoda)
G. Aquatic sow bug (Isopoda)
H. Snail (Gastropoda)

I Fingernail clam (Sphaeriidae)
J. Damselfly nymph (Zygoptera)
K. Dragonfly nymph (Anisoptera)
L. Bloodworm or midge fly larvae
   (Tendipedidae)
M. Leech (Hirundinea)
N. Sludgeworm (Tubificidae)
O. Sewage fly larvae (Psychoda)
P. Rat-tailed maggot (Tubifera-Eristalis)

Figure 12: Representative Bottom-Dwelling Macro-animals (Keup, Ingram and Mackenthun, 1966)

When pollution occurs the number of bottom-dwelling macrofauna in an area is reduced by eliminating the most sensitive ones. As the concentration of a given pollutant increases, additional species are eliminated in order of sensitivity to the pollutant until only those species that can survive the adverse conditions remain. Extreme pollution may actually eliminate all of the bottom-dwelling macrofauna in a given area. Some general tolerance patterns of bottom-dwelling macro-organisms can be identified. Stonefly nymphs, mayfly naiads, hellgrammites, and caddis fly larvae represent a grouping that is quite sensitive to environmental changes. Black fly larvae, scuds, sowbugs, snails, fingernail clams, dragonfly nymphs, damselfly nymphs, and most kinds of midge larvae have immediate intolerance. Sludgeworms, some kinds of midge larvae (bloodworms), and some leeches are tolerant to comparatively heavy loads of organic pollutants. Sewage mosquitos and rat-tailed maggots are tolerant of anaerobic environments (Keup, Ingram and Mackenthun, 1966).

The best type of survey for bottom-dwelling macrofauna is one focused on quantitative sampling to observe changes in predominance of species. The most common quantitative sampling tools are the Petersen and Ekman dredges and the Surber stream bottom or square foot sampler. Information on these samplers is in Chapter 6. The collected sample is screened with a standard sieve to concentrate the organisms; these are sorted from the retained material, and the number of each species is determined. Data are then adjusted to the number per unit area, usually to the number per square foot or meter of bottom. A professional biological interpretation is required for the implications of the species in both types and numbers found at a given sampling location (Keup, Ingram and Mackenthun, 1966).

Fish sampling is also an important component in biological monitoring. Chapter 6 contains a summary of several of the more common collecting techniques which can be used for fish sampling in a stream survey.

SELECTED REFERENCES

American Water Works Association and Water Pollution Control Federation, "Standard Methods for the Examination of Water and Wastewater", 1981, Washington, D.C.

Ballinger, D.G., "Natural Self-Purification in Surface Waters", in Training Course Manual entitled Aquatic Biology for Engineers, 1963, U.S. Public Health Service, Cincinnati, Ohio.

Baltisberger, R.J., "The Microdetermination of Mercury Species in Natural Water Systems by Liquid Chromatography", WI-222-010-75, 1975, Washington, D.C.

Baltisberger, R.J., "The Differentiation of Inorganic and Organomercury Species in Aqueous Samples", WI-221-041-76, 1977, Washington, D.C.

Bartsch, A.F. and Ingram, W.M., "Stream Life and the Pollution Environment", Public Works, Vol. 90, No. 7, July 1959, pp. 104-110.

Berman, D., Rohr, M.E. and Safferman, R.S., "Concentration of Poliovirus in Water by Molecular Filtration", EPA-600/J-80-161, 1980, U.S. Environmental Protection Agency, Cincinnati, Ohio.

Bingham, C.R. et al., "Grab Samplers for Benthic Macroinvertebrates in the Lower Mississippi River", Misc. Paper E-82-3, July 1982, U.S. Army Engineer Waterways Experiment Station, Vicksburg, Mississippi.

Bishop, C.T. et al., "Anion Exchange Method for the Determination of Plutonium in Water: Single-Laboratory Evaluation and Interlaboratory Collaborative Study", EPA/600/7-78-122, 1978, U.S. Environmental Protection Agency, Las Vegas, Nevada.

Block, J.C. and Rolland, D., "Method for Salmonella Concentration from Water at pH 3.5, Using Microfiber Glass Filters", Applied and Environmental Microbiology, Vol. 38, No. 1, July 1979, pp. 1-6.

Butler, S.S., Engineering Hydrology, 1957, Prentice-Hall, Inc., Englewood Cliffs, New Jersey, pp. 175-202.

Canter, L.W., Supplement to Environmental Impact Assessment, 1979, University of Oklahoma, Norman, Oklahoma, pp. 7-8.

Canter, L.W., "Pollutant Emissions from the Petroleum Industry", May 1981, report submitted to INTEVEP, S.A., Los Teques, Venezuela, pp. 87-89.

Cairns, J., Jr., Dickson, K.L. and Westlake, G.F., "Biological Monitoring of Water and Effluent Quality", ASTM Special Technical Publication 607, 1976, American Society for Testing Materials, Philadelphia, Pennsylvania.

Cairns, J., Jr. and Gruber, D., "A Comparison of Methods and Instrumentation of Biological Early Warning Systems", Water Resources Bulletin, Vol. 16, No. 2, Apr. 1980, pp. 261-266.

Chakravorty, R. and Van Grieken, R., "Co-precipitation with Iron Hydroxide and X-Ray Fluorescence Analysis of Trace Metals in Water", International Journal of Environmental Analytical Chemistry, Vol. 11, No. 1, 1982, pp. 67-80.

Chappelle, E.W. et al., "Rapid, Quantitative Determination of Bacteria in Water", Patent-4 385 113, 1983, U.S. Patent Office, Washington, D.C.

Chatfield, E.J., Dillon, M.J. and Stott, W.R., "Development of Improved Analytical Techniques for Determination of Asbestos in Water Samples", PB83-261651, 1983, National Technical Information Service, U.S. Department of Commerce, Springfield, Virginia.

Cook, P.P., Duvall, P.M. and Bourke, R.C., "Improved Methods of Oil-in-Water Analysis", Water and Sewage Works, Apr. 1978.

Crowther, J., "Autoclave Digestion Procedure for the Determination of Total Iron Content of Waters", Analytical Chemistry, Vol. 50, No. 4, Apr. 1978, pp. 658-659.

Crowther, J., "Semiautomated Procedure for the Determination of Low Levels of Total Manganese", Analytical Chemistry, Vol. 50, No. 8, July 1978, pp. 1041-1043.

Downes, M.T., "An Improved Hydrazine Reduction Method for the Automated Determination of Low Nitrate Levels in Freshwater", Water Research, Vol. 12, No. 9, 1978, pp. 673-675.

Elsenreich, S.J. and Hullett, D.A., "Determination of Free and Bound Fatty Acids in River Water by High Performance Liquid Chromatography", Analytical Chemistry, Vol. 51, No. 12, Oct. 1979, pp. 1953-1960.

Favretto, L., Stancher, B. and Tunis, F., "Improved Method for the Spectrophotometric Determination of Polyoxyethylene Non-Ionic Surfactants in Presence of Cationic Surfactants", International Journal of Environmental Analytical Chemistry, Vol. 14, No. 3, 1983, pp. 201-214.

Fleisher, J.M. and McFadden, R.T., "Obtaining Precise Estimates in Coliform Enumeration", Water Research, Vol. 14, No. 5, 1980, pp. 477-483.

Hellawell, J.M., Biological Surveillance of Rivers: A Biological Monitoring Handbook, 1978, Water Research Center, Stevenage, England.

Hirn, J. and Pekkanen, T.J., "The Stability of Simulated Water Samples for the Purpose of Bacteriological Quality Control", Vatten, Vol. 33, No. 3, 1977, pp. 318-323.

IHD-WHO Working Group on the Quality of Water, "Water Quality Surveys", Studies and Reports in Hydrology - 23, 1978, 350 pp., United Nations Educational, Scientific and Cultural Organization, Paris, France, and World Health Organization, Geneva, Switzerland.

Jackson, H.W., "The Interpretation of Biotic Responses to Organic Pollution", in Training Course Manual entitled Aquatic Biology for Engineers, 1963a, U.S. Public Health Service, Cincinnati, Ohio.

Jacobs, F. and Grant, G.C., "Guidelines for Zooplankton Sampling in Quantitative Baseline and Monitoring Programs", EPA/600-3-78/026, Feb. 1978, Virginia Institute of Marine Science, Gloucester Point, Virginia.

Jeffers, E.L., "Method and Apparatus for Continuous Measurement of Bacterial Content of Aqueous Samples", PAT-APPL-SN-891 247, 1978, U.S.   Patent Office, Washington, D.C.

Johnston, J.B. and Herron, J.N., "A Routine Water Monitoring Test for Mutagenic Compounds", UICU-WRL-79-0141, 1979, University of Illinois, Champaign, Illinois.

Joshi, S.R. et al., "An Indigenous Membrane Filter Medium for Enumeration of Coliform in Water", Indian Journal of Environmental Health, Vol. 20, No. 1, Jan. 1978, pp. 29-35.

Keup, L.E., Ingram, W.M. and Mackenthun, K.M., "The Role of Bottom-Dwelling Macrofauna in Water Pollution Investigations", PHS Pub. No. 999-WP-38, 1966, 23 pp., U.S. Public Health Service, Cincinnati, Ohio.

Kingsbury, R.W. and Rees, C.P., "Rapid Biological Methods for Continuous Water Quality Monitoring", Effluent and Water Treatment Journal, Vol. 18, No. 7, July 1978, pp. 319-331.

Kittrell, F.W., "A Practical Guide to Water Quality Studies of Streams", CWR-5, 1969, 135 pp., Federal Water Pollution Control Administration, U.S. Department of Interior, Washington, D.C.

Kloosterboer, J.G. and Goossen, J.T.H., "Determination of Phosphates in Natural and Waste Waters After Photochemical Decomposition and Acid Hydrolysis of Organic Phosphorus Compounds", Analytical Chemistry, Vol. 50, No. 6, May 1978, pp. 707-711.

Lantz, J.B., Davenport, R.J. and Wynveen, R.A., "Evaluation of Breadboard Electrochemical TOC/COD Analyzer: Advanced Technology Development", LSI-TR-310-4-3, 1980, Cleveland, Ohio.

Lesar, D.J. and Standridge, J.H., "Analytical Note: Sampling and Storage Methodology for the Detection of Sulfate-Reducing Bacteria", Journal of the American Water Works Association, Vol. 71, No. 7, July 1979, p. 406.

Manahan, S.E. et al., "An Analytical Method for Total Heavy Metal Complexing Agents in Water and Its Application to Water Quality Studies", W74-02658, 1973, University of Missouri, Columbia, Missouri.

Manja, K.S., Maurya, M.S. and Rao, K.M, "A Simple Field Test for the Detection of Faecal Pollution in Drinking Water", Bulletin of the World Health Organization, Vol. 60, No. 5, 1982, pp. 797-801.

Manning, D.C., "Water/Wastewater Analysis by Atomic Absorption Spectroscopy", Water and Sewage Works, 1979.

Marks, A., "Ion Selective Electrodes: Progress and Potential", Intech, Vol. 29, No. 6, 1982, pp. 9-10.

Martins, M.T., Alves, M.N. and Sanchez, P.S., "Comparison of Methods for Pseudomonas Aeruginosa Recoveries from Water", Environmental Technology Letter, Vol. 3, No. 9, 1982, pp. 405-410.

Matsumoto, G., "Comparative Study on Organic Constituents in Polluted and Unpolluted Inland Aquatic Environments II.   Features of Fatty Acids for Polluted and Unpolluted Waters", Water Resources, Vol. 15, No. 7, 1981, pp. 779-787.

National Environmental Research Center, "Methods for Chemical Analysis of Water and Wastes", EPA-16020-07/71, 1971, U.S. Environmental Protection Agency, Cincinnati, Ohio.

National Technical Information Service, "Fresh Water Mussels and Water Quality. 1970. June, 1983 (Citations from the NTIS Data Base)", 1983, U.S. Department of Commerce, Springfield, Virginia.

Nemerow, N.L., Scientific Stream Pollution Analysis, 1974, McGraw-Hill Book Company, New York, New York, pp. 6-19.

Reasoner, D.J., Blannon, J.C. and Geldreich, E.E., "Rapid Seven-Hour Fecal Coliform Test", Applied and Environmental Microbiology, Vol. 38, No. 2, Aug. 1979, pp. 229-236.

Rock, J.C. and Post, J.A., "Field Analysis of Ten Parts Per Billion Trichloroethylene in Water, Using a Portable, Self-Contained Gas Chromatograph", 14th Annual Conference on Trace Substances in Environmental Health, 1980, University of Missouri, Columbia, Missouri.

Roline, R.A. and Miyahara, V.S., "Evaluation of the Algal Assay Bottle Test", REC-ERC-80-1, 1979, U.S. Bureau of Reclamation, Denver, Colorado.

Salim, R. and Cooksey, B.G., "Effect of Centrifugation on the Suspended Particles of River Waters", Water Resources, Vol. 15, No. 7, 1981, pp. 835-839.

Schillinger, J.E., Evans, T.M. and Stuart, D.G., "Rapid Determination of Bacteriological Water Quality by Using Limulus Lysate", Applied and Environmental Microbiology, Vol. 35, No. 2, Feb. 1978, pp. 376-382.

Seitz, W.R., "Evaluation of Flame Emission Determination of Phosphorus in Water", EPA-660/2-73-007, 1973, U.S. Environmental Protection Agency, Corvallis, Oregon.

Sekerka, I. and Lechner, J.F., "Potentiometric Determination of Organohalides in Natural Water Using Tenax Adsorption and Combustion", International Journal of Environmental Analytical Chemistry, Vol. 11, No. 1, 1982, pp. 43-52.

Sharma, S.R., Rathore, H.S. and Ahmed, S.R., "New Specific Spot Test for the Detection of Malathion in Water", Water Resources, Vol. 17, No. 4, 1983, pp. 471-473.

Slabbert, J.L. and Morgan, W.S.G., "Bioassay Technique Using Tetrahymena Pyriformis for the Rapid Assessment of Toxicants in Water", Water Resources, Vol. 16, No. 5, 1982, pp. 517-523.

Slooff, W., "Biological Monitoring Based on Fish Respiration for Continuous Water Quality Control", Second International Symposium on Aquatic Pollutants, 1977, National Institute for Water Supply, Amsterdam, Netherlands.

States, J.B. et al., "A Systems Approach to Ecological Baseline Studies", FWS/OBS-78/21, Mar. 1978, U.S. Fish and Wildlife Service, Fort Collins, Colorado.

Stepanenko, V.E. and Muslova, N.M., "Chromatographic Determination of Organic Carbon in Aqueous Solutions", Zavodskaya Laboratoriya, Vol. 44, No. 9, Sept. 1978, pp. 1068-1071.

Stout, G.E. et al., "Baseline Data Requirements for Assessing Environmental Impact", IIEQ-78-05, May 1978, Institute for Environmental Studies, University of Illinois, Urbana-Champaign, Illinois.

Thurnau, R.C., "Ion Selective Electrodes in Water Quality Analysis", EPA/600/2-78-106, 1978, U.S. Environmental Protection Agency, Cincinnati, Ohio.

Ullman, F.G., "Investigation of Laser Raman Spectroscopy for Analysis of Water Quality", W77-01817, 1976, U.S. Department of the Interior, Washington, D.C.

Ward, D.V., Biological Environmental Impact Studies: Theory and Methods, 1978, Academic Press, New York, New York.

Water Research Centre, "Simultaneous Multi-Element Analysis of Aqueous Solutions", WRL-RN-8, 1977, Stevenage, England.

Weber, C.I., "Biological Field and Laboratory Methods for Measuring the Quality of Surface Waters and Effluents", EPA-670/4-73-001, 1973, U.S. Environmental Protection Agency, Cincinnati, Ohio.

Weiss, C.M., "Evaluation of the Algal Assay Procedure", EPA/600/3-76-064, 1976, U.S. Environmental Protection Agency, Corvallis, Oregon.

CHAPTER 5

LOCATION OF SAMPLING STATIONS AND
FREQUENCY OF SAMPLING

Two of the most important decisions related to planning a sampling program for a river are associated with the location of sampling stations and the frequency of sampling. Stations should be located such that proper samples taken from the areas would be representative of the entire flow of the stream at that point and at that time. Sampling stations should not be established at locations where mixing is incomplete, or where significant differences in water composition exist in the stream cross section (Brown, Skougstad and Fishman, 1970). An ideal sampling station in terms of water quality would be a cross section of a stream at which samples from all points on the cross section would yield the same concentrations of all constituents, and a sample taken at any time would yield the same concentrations as one taken at any other time (Kittrell, 1969). It is obvious that ideal locations for sampling stations are extremely limited, if not possible to identify. Accordingly, the selection of a sampling station location has to be based on a compromise from the ideal to what exists in the area. This chapter begins with a background discussion of the concept of representative sampling and is followed by general information on key factors in sampling station selection. Unique requirements related to biological sampling station locations are also presented along with the concept of having basic stations to be supplemented with auxiliary stations for specific purposes in the overall conduct of a sampling study.

Rules for frequency and duration of sampling cannot be precisely defined to meet every situation. A key aspect related to sampling frequency is that better plans can be made, with a greater statistical basis, as more information becomes available. In other words, following preliminary gathering of information on the study river, greater statistically based sampling frequencies can be determined. Therefore, this chapter will delineate some general principles for determining sampling frequency as well as provide information related to numbers of measurements.

DESIRABILITY OF REPRESENTATIVE SAMPLING

The homogeneity of a stream throughout a given cross section is determined by such physical factors as turbulence and distance from large inflows. At and immediately below the confluence of tributary streams there is frequently a distinct physical separation between the waters of the tributary and the main stream. This effect is more pronounced, and may persist for many miles downstream, if the composition of the two waters differ significantly with regard to temperature, and/or to dissolved and suspended solids. Theoretically, a sample representing the overall composition of a stream can be obtained by compositing several depth-integrated samples of equal volume taken at places with equal flow across the cross section. Due to the large number of samples required, it is highly desirable to find a stream section in which the water composition is uniform with depth and cross section. Such sections can usually be located on small and medium-size streams, but are frequently impossible to find on large rivers. Where uniform sections can be found, however, the sampling procedure may often be simplified to the extent that a single "grab" sample may be obtained that is representative of the stream composition (Brown, Skougstad and Fishman, 1970).

One of the things which must be determined for a river study is whether or not the initially selected sampling station locations represent points that are uniform in terms of water quality constituents both in depth and cross section. If this is not the case, then consideration should possibly be given to relocation of the stations. Information on general considerations for establishing the homogeneity of stream cross sections is contained in Appendix H (Sanders, 1980).

SITE SELECTION

Many factors are involved in proper selection of sampling station locations. These include the objectives of the stream study; accessibility of the site; flows, mixing, and other physical characteristics of the water body; point and diffuse sources of contamination; physical structures such as dams, weirs, and wing walls; and available personnel and facilities. As indicated earlier, actual identification of a sampling station location represents a compromise between the above factors.

The study objectives should be considered in the identification of sampling station locations. If the objective is limited to establishing baseline water quality, then a particular set of sampling station locations might be appropriate. On the other hand, if additional objectives are included, such as to determine the effects of existing pollution sources on the river quality, then other sampling station locations may be necessary. Additionally, if the objective is to establish the waste assimilative capacity for the river, then still different sampling station locations would be possible. Appendix H contains some discussion related to considering the objectives of the sampling program in sampling station location (Sanders, 1980).

A practical factor for consideration in sampling station location is the accessibility to the location. Bridges are a logical choice for locating a stream sampling station because they provide ready access and also permit sampling at any point across the stream (U.S. Geological Survey, 1977). Station locations can be chosen without regard to other means of access if the stream is navigable by an available boat. A combination of bridges and boats may prove to be the best system in some situations. Walking to collect samples may be feasible in a few cases, but this method usually should be chosen for only very small streams located near highways or access roads. Sampling by helicopter has advantages of ready access, speed, and minimum of physical effort (Kittrell, 1969). However, this type of sampling is expensive and may not be justified in a given sampling program.

Establishment of stations at marked changes in physical characteristics of the stream channel is desirable. For example, a stream reach between two adjacent stations should not include both a long rapid section of swift, shallow water with a rocky bottom, and a long section of deep, slow-moving water with a muddy bottom. Stations at each end of the combined reach would yield data on certain rates of change, such as reaeration, that would be an unrealistic average of two widely different rates. Much more would be learned of the actual natural purification characteristics of the stream by the insertion of a third station within the reach between the rapids and quiet water sections (Kittrell, 1969).

Other factors which must be considered are the locations of point and nonpoint sources of municipal and industrial waste discharges as well as agricultural drainage. This is particularly important in terms of establishing causative factors for various water quality characteristics. If it is desired to locate sampling stations to reflect these inflow influences, it will be necessary to locate them sufficient distances downstream to allow for homogeneous mixing to occur (U.S. Geological Survey, 1977).

An additional aspect of inflow into a river is associated with tributary streams. The sampling of every tributary stream on a river would probably not be feasible. A tributary with a flow that is less than 10 to 20 percent of the main stream flow would probably not need to be sampled unless it represents a stream with unique water quality characteristics which might be resulting from being polluted (Kittrell, 1969). The sampling station on a tributary should be as near the mouth as is feasible. Frequently, the mouths of tributaries may be entered from the main stream when sample collection in the main stream is by boat. Care should be exercised to avoid collecting water from the main stream that may flow into the mouth of the tributary on either the surface or bottom because of differences in density resulting from temperature, dissolved salts, or turbidity.

Artificial physical structures such as dams, weirs, and wing walls can cause changes in hydraulic flow patterns and thus influence the representativeness of the water quality at specific locations. If there are artificial physical structures along the study river, these should be considered in terms of locating sampling stations. Artificial structures usually create quiet, deep pools in river reaches that, by comparison, formerly were swift and shallow (Kittrell, 1969).

Another practical consideration in sampling station location is associated with the personnel and facilities assigned to the study. An over-enthusiastic number of sampling stations can literally exceed the abilities of the study staff to sample and conduct the necessary analytical work. Careful consideration should be given to developing a sampling and analysis program that can be handled within the capabilities of the assigned staff and laboratory facilities.

BIOLOGICAL STATIONS

Considerations in sample station selection discussed thus far have been directed toward sampling for chemical and bacterial constituents. Some of the same factors certainly apply in the case of biological sampling, but additional factors must also be considered. Chemical characteristics of the water have an impact on the aquatic organisms in a stream, and biological reactions in turn affect chemical characteristics. This has been discussed in Chapter 4. Full understanding and interpretation of both biological and chemical findings requires consideration of these mutual interactions; therefore, if at all possible, biological and chemical samples should be collected at or near the same stream locations (Kittrell, 1969).

Plankton samples should be collected at the same stations and in essentially the same way as samples for chemical analyses. They should be collected within a foot of the surface, since some organisms tend to congregate near the surface. Samples of benthic organisms, on the other hand, frequently should not be collected at the same stations. Sampling from bridges across a stream for bottom organisms may not be desirable since the stream bottom in the vicinity of bridges may have been disturbed due to human activities, and the types of sampling equipment used are not water tight, thus water leaking from the equipment as it is hauled up to bridges may carry away many organisms and thus yield a nonrepresentative sample.

The number of bottom sampling points on a cross section should vary with the width of the stream. Generally, one point is adequate for a stream up to 20 feet wide, two points between 20 and 150 feet, and three points over 150 feet (Kittrell, 1969). Location of bottom organism sampling stations where vertical mixing of wastes or tributary streams is complete is essential since the wastes or other tributary influences must reach the bottom if their effects are to be detected. Vertical mixing usually occurs in such short distances that lack of vertical mixing only rarely constitutes a problem in biological station location. Multi-point sampling of bottom organisms is necessary on a cross section where lateral mixing is incomplete, just as it is frequently for chemical and bacterial data. Bottom organism data from a cross section should not be averaged as they usually are for chemical and bacterial data; however, the biological data for each point in the cross section should be reported and discussed separately with emphasis given to the locations that reflect the greatest effects of the waste materials (Kittrell, 1969).

BASIC AND AUXILIARY STATIONS

One of the possible options in sampling station location is to identify basic stations as well as a series of auxiliary stations (IHD-WHO Working Group on the Quality of Water, 1978). Basic stations are used for classifying water resources, collection of baseline data, and determination of water quality. Such stations are usually sparsely scattered over the entire basin. Appendix H provides information on both the macro as well as micro location of basic sampling stations (Sanders, 1980). Basic stations are usually located at the mouth of main streams and principle tributaries, downstream of river development projects, above and downstream of waste outfalls and industrial and urban centers, and at points of major water withdrawal.

Investigations on the effects of pollutants discharged into a stream, determination of assimilative capacity, and other special studies may require auxiliary stations (IHD-WHO Working Group on the Quality of Water, 1978). These stations are purposely related to each other and may be moved to another place or operated only temporarily. The spacing of auxiliary stations is generally much closer than the spacing of basic stations. Auxiliary stations are more typically located on the basis of time of water travel from one point to another. In general, when monitoring dissolved oxygen, biochemical oxygen demand and decomposing chemical impurities downstream from waste discharges, stations are spaced initially at one-half days time-of-water-travel below waste outfalls for a flow period of two or three days, and then at about one-day intervals for another three to four days or the remaining length of the study reach. If the survival of coliform or other indicator organisms is studied, the same pattern may be followed, but the observations may be extended to a total time of flow of 10 to 20 days (IHD-WHO Working Group on the Quality of Water, 1978).

PRINCIPLES FOR DETERMINING SAMPLING FREQUENCY

Sampling frequency depends largely on the purpose of the sampling network, the relative importance of the station, and the anticipated variability of the water quality data at the station. Sampling frequency at a particular station will also be influenced by the accessibility of the station as well as the personnel and associated laboratory facilities available for the study (IHD-WHO Working Group on the Quality of Water, 1978). An additional factor related to data variability is associated with the precipitation or anticipated river flows in a given area. It is entirely appropriate to consider a sampling program that would focus on both wet and dry seasons of the year.

One approach for determining sampling frequency is to consider whether the stations are basic or auxiliary stations. At newly initiated basic stations the samples may be collected at a higher frequency so that within two to three years a sufficient number of observations are available for a statistical evaluation of variations, cycles and trends (IHD-WHO Working Group on the Quality of Water, 1978). After an initial period of relatively high frequency sampling, the frequency can either be increased or decreased for the longer time period. For basic station networks operated for collection of baseline data, the sampling frequency may be from three to four times per year to monthly collection of samples. In any event, no less than one sample should be taken in each season if there are four pronounced hydrological and climatological seasons, or no less than one sample should be taken in each of two seasons if the characteristics can be classified in terms of wet and dry periods. In addition to specifically planned sampling periods, it may be desirable to collect two or three random samples per year at a given basic sampling station.

Auxiliary stations are usually operated for shorter periods of time and for more specific purposes than a basic sampling network for establishing baseline water quality. Accordingly, the frequency of sampling at auxiliary stations may be considerably higher than for basic sampling stations. For example, if an auxiliary station is to be operated for a period of several weeks, the number of samples collected should be enough to give a sufficiently high reliance to the mean of observations. It is suggested that a total number of 18 to 24 samples are required for meaningful statistical analysis (IHD-WHO Working Group on the Quality of Water, 1978).

NUMBER OF SAMPLES

One of the basic principles for consideration in sampling frequency identification is that increasing the number of samples will reduce the standard error of the mean value. However, since the standard error of the mean varies inversely as the square root of the number of observations, there will only be a rather small gain by increasing the number of samples beyond the numbers as suggested above. For example, the precision yielded by 16 samples will be increased only 20 percent by increasing the number of samples to 25, with the additional 9 samples representing a 56 percent increase in the overall number of samples (Kittrell, 1969).

Experience has shown that, as a rule of thumb, 20 to 25 samples collected at a station during a period of relatively stable waste discharge, stream flow and temperature yield reasonably representative mean Most Probable Number

(MPN) values for coliform bacteria (Kittrell, 1969). Fewer samples may yield acceptable mean values for other constituents since the method for MPN estimation is relatively inprecise.

Another consideration relative to sampling frequency is that it would be undesirable to average the results of samples taken over a wide range of stream flows. Likewise, results on samples taken during hot weather and cold weather, or other periods of large temperature differences, should not be included in a common average. Data collected throughout the year or for several years may be presented in terms of averages, and perhaps maxima and minima, for each month or for each seasonal period or wet or dry season.

SELECTED REFERENCES

Brown, E., Skougstad, M.W. and Fishman, M.J., "Techniques of Water Resources Investigations of the United States Geological Survey", Chapter A1--Methods for Collection and Analysis of Water Samples for Dissolved Minerals and Gases, Book 5--Laboratory Analysis, 1970, 160 pp., U.S. Geological Survey, Washington, D.C.

IHD-WHO Working Group on the Quality of Water, "Water Quality Surveys", Studies and Reports in Hydrology - 23, 1978, 350 pp., United Nations Educational, Scientific and Cultural Organization, Paris, France, and World Health Organization, Geneva, Switzerland.

Kittrell, F.W., "A Practical Guide to Water Quality Studies of Streams", CWR-5, 1969, 135 pp., Federal Water Pollution Control Administration, U.S. Department of Interior, Washington, D.C.

Sanders, T.G., editor, "Principles of Network Design for Water Quality Monitoring", July 1980, 312 pp., Colorado State University, Ft. Collins, Colorado.

U.S. Geological Survey, "National Handbook of Recommended Methods for Water-Data Acquisition", Ch. 5--Chemical and Physical Quality of Water and Sediment, Jan. 1977, pp. 5-7 through 5-16c, Reston, Virginia.

CHAPTER 6

SAMPLE COLLECTION AND ANALYSES

Another important element in planning a sampling program for a river is associated with the collection, identification and analyses of samples. It is important that the water, sediment and biota samples be collected in a satisfactory manner using appropriate containers and equipment. It is also necessary that the samples be identified and the entire process conducted in a safe manner. Provision of adequate analytical laboratory services is also required, including consideration of general laboratory planning, mobile laboratories, and laboratory safety. This chapter will address these issues.

SAMPLING EQUIPMENT

A wide range of sampling equipment is available for usage in water quality surveys. One aspect of a baseline study includes selection of appropriate containers as well as sampling equipment for water, plankton, sediment, and fish. Factors pertinent in selecting containers for collection and storage of water samples are resistance to solution and breakage, efficiency of closure, size, shape, weight, availability, and cost (Brown, Skougstad and Fishman, 1970). Preferences for one type of container over another are varied, and selection is guided largely by experience of the individuals doing the purchases as well as availability of the containers. Although a great number of glass bottles are still in use, the current trend is toward the use of polyethylene, teflon, or other plastic containers.

Water samples for physical, chemical and biological analyses are taken in much the same manner. For bacteriological analysis it may be necessary to use sterilized equipment. One to three liters of sample should suffice for most physical, chemical and biological analyses (IHD-WHO Working Group on the Quality of Water, 1978). The simplest form of a water sampling device is a bottle attached to a string. To lower a plastic or glass bottle into a body of water it is necessary to use a bracket or holder of sufficient weight to overcome the bouyancy of the bottle and allow it to sink as rapidly as desired. An example of a sample bottle holder for manual sampling is contained in Figure 13 (IHD-WHO Working Group on the Quality of Water, 1978).

When water from a particular depth is to be collected, several types of sampling devices are available. For good accuracy and for water depths greater than 50 meters, the Kemmerer sampler represents one choice, with a diagram of this sampler shown in Figure 14 (IHD-WHO Working Group on the Quality of Water, 1978). The Kemmerer sampler, as well as several others of this type, consists of a bottle, open at both ends, which is lowered to the desired depth in the open position in order to allow water to stream through the bottle. Closure is achieved by a drop weight or a messenger which slides down the supporting wire or cord. In this way, a sample of water is isolated and is not affected when the bottle is pulled up.

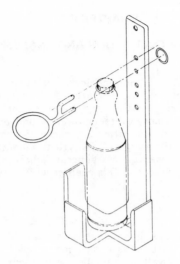

Figure 13: Sample Bottle Holder for Manual Sampling (IHD-WHO Working
Group on the Quality of Water, 1978)

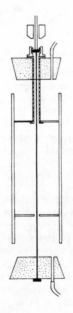

Figure 14: Kemmerer Sampler (IHD-WHO Working Group on the Quality of
Water, 1978)

Water sampling can also be accomplished by the use of continuous water quality monitors. The following information is presented to indicate some of the general factors associated with planning for usage of continuous monitoring systems.

One of the first limitations which needs to be recognized in continuous water quality monitoring is that not all water quality parameters can be measured using these systems. Table 23 identifies relevant characteristics that can be measured reliably with adequate accuracy (IHD-WHO Working Group on the Quality of Water, 1978). An additional consideration in the use of water quality monitors is the cost of a permanent station for measuring these parameters. A conservative estimate is that the installation of one water quality monitor for measuring the characteristics identified in Table 23 would be in excess of $25,000.

Table 23: Accuracy of Automatic Water Quality Monitors (IHD-WHO Working Group on the Quality of Water, 1978)

Data	Unit	Range	Accuracy
Temperature	$°C$	-10 to 40	$\pm 0.5°C$ (linear scale)
Dissolved oxygen	% of air saturation value	0–100 0–200	$\pm 1\%$ of saturation $\pm 2\%$ of saturation (linear scale)
Ammoniacal nitrogen	mg N/l	0–5 0–10 0–50	$\pm 5\%$ of reading (log scale)
Organic matter*	mg C/l	0–10 0–100	$\pm 5\%$ of reading (log scale)
Suspended matter	mg/l based on formazin standard	5–500 50–5,000	$\pm 5\%$ of reading (log scale)
Conductivity	$\mu s/cm^2$	5–5,000 50–50,000	$\pm 5\%$ of reading (log scale)
Chloride	mg Cl/l	0.5–500 25–25,000	$\pm 5\%$ of reading (log scale)
Hardness	mg/l as $CaCO_3$	10–1,000	$\pm 5\%$ of reading (log scale)
pH value	units	2–11	± 0.1 (linear)
Sunlight intensity	cal/cm² h	0–120	± 1.2 (linear)
Dissolved carbon dioxide	mg/l as CO_2	0–100	$\pm 5\%$ of reading (log scale)

* In terms of absorption of ultra-violet light.

An additional issue of importance in the use of continuous monitoring systems is associated with data generation and transmission. A potential suitable system for data recording would be associated with recording in serial binary or binary-coded decimal form on magnetic tape using battery-powered incremental recorders. A more sophisticated level of data transmission would involve the use of telemetry systems for relaying data to one or more master stations. The cost associated with data recording and telemetry would also be high (IHD-WHO Working Group on the Quality of Water, 1978). These comments on the use of continuous water quality monitoring systems are not intended to be negative. There may be occasion where these systems could be appropriately used in the continuing sampling program on a river. Some possible purposes of the use of systems such as these are as follows (IHD-WHO Working Group on the Quality of Water, 1978):

(1)   To give warning of unusual events such as unexpectedly high concentrations of organic pollution;

(2)   To provide a better basis for definition of cyclical fluctuation and characteristic sequences of events than would be afforded by the sampling program; and

(3)   To provide a means for checking the relations assumed between sampling frequency and precision of estimates.

Table 24 contains key comments on 19 references related to either automatic sampling or the use of remote sensing in water quality surveys. Appendix I contains the abstracts for the cited references. Growing usage is being made of remote sensing in river water quality studies.

Table 24: References on Automatic Sampling and Remote Sensing

Authors (Year)	Comments
Briggs, Page and Schofield (1977)	Sensors in continuous water quality monitors.
Brinkhoff (1977)	Data collection and presentation with continuous on-site monitors.
Cavagnaro (1977)	154 abstracts on continuous water quality monitoring.
Cavagnaro (1978)	216 abstracts on continuous water quality monitoring.
Cavagnaro (Vol. 2, 1979)	190 abstracts on continuous water quality monitoring.
Cavagnaro (Vol. 3, 1979)	138 abstracts on continuous water quality monitoring.
Cavagnaro and Hundemann (1980)	168 abstracts on continuous water quality monitoring.
Davis (1973)	Case study of reliability of continuous water quality monitoring.
Eagleson, Morgan and McCollough (1978)	Automated biomonitor and satellite data transmission.
Fisher and Siebert (1977)	Automatic sample collection as a function of stream gage height.
Grana and Haynes (1977)	Remote water quality monitoring system.

Table 24: (continued)

Authors (Year)	Comments
Koehler (1978)	Automated water sampler responsive to stream hydraulic changes.
Lillesand, Scarpace and Clapp (1973)	Use of color infrared photography in water quality monitoring.
National Technical Information Service (1982)	297 abstracts on continuous water quality monitoring, including the use of satellite remote sensing.
Rogers (1975)	Use of Landsat for surveillance of lake water quality.
Thiruvengadachari et al. (1983)	Considerations in collection of ground truth information for remote sensing.
Wallace, Lovelady and Ferguson (1981)	Remotely deployable water quality monitoring system.
Welby (1976)	Multispectral photography for water quality monitoring.

Water collected through any of the methods of sampling discussed above can also be used for measurement of planktonic organisms. Besides these there are other samplers which combine the operation of sampling and separation of plankton. One such example is the Juday plankton trap which samples five liters of water. A picture of this device is in Figure 15 (IHD-WHO Working Group on the Quality of Water, 1978). When using the Juday plankton trap, the sampler is lowered to the depth to be sampled and a messenger is sent down to close the top and bottom panels. The sampler is hauled to the surface, preferably using a hoist, and taken out of the water carefully, allowing the contained five liters of water to be drained through the plankton bucket. Following filtration, the bucket is removed and its contents transferred to a sample bottle using distilled water if necessary. This apparatus will take large quantitative samples very satisfactorily, however, it can be awkward to handle in a boat with rough waters in existence.

Another method of plankton sampling involves using a plankton net. Some plankton nets have had flow measuring, opening and closing devices added for operation from a moving boat. When these devices are utilized, quantitative samples can be obtained. If only qualitative samples are required, an ordinary plankton net and bucket operated either from a boat or from shore is all that is necessary. A representative collection of the plankton present in a concentrated form for species composition can be obtained using a plankton net, but it will not provide a quantitative measure.

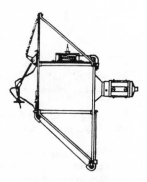

Figure 15:   Juday Plankton Trap (IHD-WHO Working Group on the Quality of
Water, 1978)

Sampling equipment for collection of sediment samples as well as bottom-dwelling macrofauna include several types of dredges and related equipment. Quantitative results can be obtained from samples collected with these devices. Three examples include the Eckman dredge, the Petersen dredge, and the Surber sampler. The Eckman dredge shown in Figure 16 is a fairly light, spring-triggered, brass dredge made in several sizes (about 15 cm x 15 cm, 24 cm x 24 cm, and 30 cm x 30 cm; the 15 cm x 15 cm size is the most common). After reaching the bottom a messenger is sent down the line to trigger closure of the jaws and enclose a sample of bottom sediment as well as associated bottom-dwelling macrofauna (IHD-WHO Working Group on the Quality of Water, 1978).

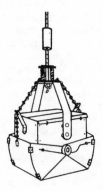

Figure 16:   Eckman Dredge (IHD-WHO Working Group on the Quality of
Water, 1978)

The Petersen dredge shown in Figure 17 is heavier than the Eckman dredge. The Petersen dredge is not spring equipped, requires no messenger, and its closing is induced by the release of tension in the line once the dredge has settled on the stream bottom.

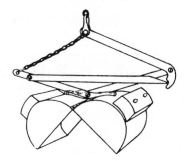

Figure 17:   Petersen Dredge (IHD-WHO Working Group on the Quality of
Water, 1978)

The Surber sampler shown in Figure 18 is a light-weight device for
procuring biological samples in water depths up to 0.75 m for fast flowing
streams (IHD-WHO Working Group on the Quality of Water, 1978). The sampler
consists of a strong, close-woven fabric approximately 70 cm long. This net is
held open by a one sq ft metal frame hinged at one side to another frame of
equal size. In operation, the frame which supports the net is in a vertical
position while the other frame is locked in a horizontal position against the
bottom. The net opening faces upstream. Within the framed area, rocks and
other bottom deposits are dug to a depth of at least 6 cm. The dislodged
organisms then drift into the open net and comprise the sample.

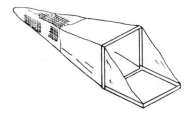

Figure 18:   Surber Sampler (IHD-WHO Working Group on the Quality of
Water, 1978)

Some common techniques for collecting fish samples are summarized in
Table 25 (Jackson, 1963b). The techniques include the use of seines, nets, traps,
trawls and poisoning.

SAMPLING DEPTH

Another consideration in the planning of a monitoring program for a river
is the depth at which samples should be collected. At stations where complete
mixing of waste inputs or tributary streams has taken place only one sample
may be needed from the midpoint of the depth. For chemical analyses, samples
may be collected at mid-depth or 1.5 m from the surface, whichever is less.
For bacteriological analyses, samples should be collected 0.5 m below the water

Table 25: Common Fish Collecting Techniques (Jackson, 1963b)

Technique	Comments
Seines	Straight seines range from 4 or 6 to upwards of 50 ft in length. "Common Sense" minnow seines with approximately 1/4 in mesh are widely used along shore collecting of the smaller fishes. Bag seines have an extra trap or bag tied in the middle which helps trap and hold fishes when seining in difficult situations.
Gill Nets	Used in offshore and deep waters. They range in length from approximately 50 yds upward, and of a mesh size designed to catch a specified size range of fish. The trammel net is a variation of the gill net.
Traps	Range from small wire boxes or cylinders with inverted cone entrances to semi permanent weirs a half mile or more in length. All tend to induce fish to swim into an inner chamber protected by an inverted cone or V-shaped notch to prevent escape.
Otter Trawls	Submarine seines or traps held open by two doors or "otter boards" and dragged by a power boat. Formerly used only by marine fisheries, they are now employed in fresh waters.
Electric Seines	Widely employed by fishery workers in small and difficult streams. They may also be used in lakes under certain circumstances. Electric screens are much used at dams and raceways to keep fish from entering certain areas.
Poisoning	Most widely used and most generally satisfactory is rotenone in various formulations, although many others have been employed from time to time. With suitable technique, all the fish may be collected from a section of stream or the arm of a lake only, or the entire fish population may be killed. Under suitable circumstances, fish may even be killed selectively according to species, such as the removal of gizzard shad by 1 ppm of rotenone without harming other species.

surface (IHD-WHO Working Group on the Quality of Water, 1978). It may be necessary to consider sampling at several depths near the mouth of a river, particularly in those areas that might be influenced by stratification due to the salt water wedge.   This can best be determined following the collection of preliminary information.

FIELD NOTES

The sample collector should label all samples and complete all field analyses and necessary records before leaving a sampling station (Kittrell, 1969). Any attempt to depend on memory from one station to the next, or until the end of the sampling run is certain to lead to undesirable results.   The identification associated with each sample should include the following information: date of sampling, station identification, time of sample collection, depth of sample, water temperature, water pH (if determined), condition of weather (sunny, cloudy, light rain), appearance of stream (clear, turbid, oil, floating debris), gage reading (if any), and any other special observations that might be useful in subsequent interpretation of water quality information. A record should also be made of the sample collection crew.

SAFETY PRECAUTIONS

Sampling of a river will likely be done under a wide range of conditions, and the work may have certain hazards. A knowledge of the hazards involved and the means by which they can be minimized should be helpful in preventing accidents and in providing greater safety for persons involved in sample collection (Rainwater and Thatcher, 1960).   Surface water samples may be collected by wading or from bridges or boats. Wading is one of the easiest methods to collect samples from many streams. Rubber boots or breast-high waders should be standard equipment.   A wading rod or similar probing instrument is essential to safe wading. By probing ahead, the collector can estimate the current and locate holes and even quicksand. A general rule of thumb is that wading should not be attempted if the depth of the water in feet multiplied by the water velocity in feet per second equals ten or more, but this criterion must be modified by many factors peculiar to the site and to the season of the year (Rainwater and Thatcher, 1960).

When sampling from bridges, traffic using the bridge is the most serious hazard. Sometimes bridges have walkways for pedestrian traffic, or catwalks suspended at the side or beneath the bridge; however, more often than not the collector must work in the traffic lanes. If necessary to interfere with traffic, suitable arrangements with local authorities should be made in advance (Rainwater and Thatcher, 1960).

The majority of the water samples to be collected in a given river survey will probably be collected from boats. The degree of hazard involved is generally dependent on the selection and operation of the boat. Lifejackets would represent essential equipment for each person to be involved in the sampling crew as well as the boat operator. Boat operations on streams during floods is particularly hazardous, and, where measurements by boat are required, the services and equipment of skilled local personnel should be utilized for boat operation. The boat operator must be alert for shallows and for floating and submerged drift materials. A good boat operator should also check items such as oars, life preservers, buoyancy tanks, lights, motor performance, gasoline supply and spare parts before the boat is used for sampling (Rainwater and Thatcher, 1960).

One of the items which should be provided for each sampling crew is a standard first-aid kit. First-aid represents the immediate and temporary care of an injured or suddenly ill person until the services of a qualified person can be obtained. The application of first aid is not a substitute for competent medical attention (Rainwater and Thatcher, 1960).

GENERAL LABORATORY PLANNING

Traditional planning for an analytical laboratory involves the placement of work benches around the walls and then, if space is available, an island of benches placed in the middle of the room (IHD-WHO Working Group on the Quality of Water, 1978). This arrangement is somewhat inefficient because the corner areas are usually inaccessible and the island of benches may be difficult to service with water, electricity, gas, and other utilities.   An alternative laboratory design is to use a peninsular structure where the services can be more easily piped in from a central core. Figure 19 illustrates a modular design based on the peninsular structure. Space arrangement in the module shows adequate bench space for two workers, and since the services run under the windows there is no problem in reaching any point on the bench. A minimum adequate bench width is 60 cm for a single-sided bench, and 120 cm for a double-sided bench, especially if the surface is clear of projections.

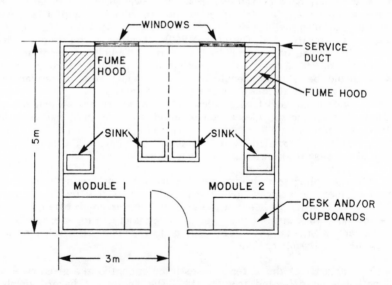

Figure 19:   Example of Two-Module Laboratory (IHD-WHO Working Group on the Quality of Water, 1978)

Laboratory space should be organized to allow an easy flow of samples from the first test bench to the final test point and wash-up area (IHD-WHO Working Group on the Quality of Water, 1978).   The washing facilities, especially those for sterile techniques must be carefully considered.   They should be separate from the main laboratory area and should be equipped with acid handling facilities and acid resistant plumbing, as should all plumbing in a laboratory. The laboratory space should be free of vibration, particularly in the analytical balance area. Concrete microscope benches may also be essential for critical work.

MOBILE LABORATORY PLANNING

A mobile laboratory can be useful for the gathering of samples and preliminary analytical work in large field studies. It can also allow site specific studies as well as special problem chemical studies to be conducted in an easier fashion. Mobile units should not be considered as alternatives to a central analytical laboratory, but rather an extension of the capacity and usefulness of a permanent central laboratory facility (IHD-WHO Working Group on the Quality of Water, 1978). Mobile laboratories can be built on conveyances such as vans, trucks, or towed trailers. In designing the mobile laboratory layout, the analyses which will be performed should be identified. Routine analyses in baseline water quality monitoring studies may include specific conductance, color, hardness, pH, temperature, alkalinity, and dissolved oxygen plus others as appropriate. The remainder of the analytical work can be performed at the central laboratory.

Figure 20 contains the layout of a mobile laboratory that is 5 m long, 2.5 m wide, and built into a towed trailer or a van (IHD-WHO Working Group on the Quality of Water, 1978). This laboratory can be towed with a single axle truck. The utilities should include a 110 or 220 volt AC system with a water-cooled generator operating a roof-mounted air conditioner, a small refrigerator operating on electricity or on propane gas, a propane gas range, a vacuum pump, an incubator, fluorescent lights, a water heater, an exhaust fan, an exhaust pump, a blower heating system, gas cylinders, perhaps a drying oven and distilling apparatus, fresh water storage tank, waste water storage tank, external hook-up for 110 or 220 volt AC, and an external hook-up for fresh water. The plumbing should not be galvanized if biological analyses are performed occasionally. The work area should be resistant to acids and alkalies, and there should be adequate ventilation and window space.

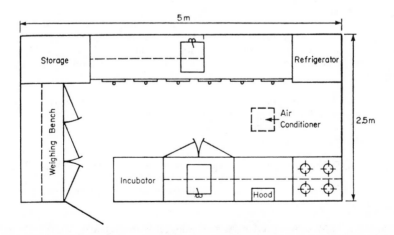

Figure 20:   Layout of a Mobile Laboratory (IHD-WHO Working Group on the Quality of Water, 1978)

LABORATORY SAFETY

A common issue of concern in analytical work, whether in a central facility or a mobile unit, is related to laboratory safety. Personal safety equipment should be worn, and there should be emergency equipment for flammable and toxic materials such as automatic sprinklers, showers, fire extinguishers, and eye wash units. The laboratory facility should be constructed of noncombustible materials and the basic furnishings should be acid and alkali resistant. Laboratories must also have sufficient air input to provide proper air movement through the exhaust system. Ventilated hoods are important for handling toxic chemicals and to provide a place for some routine analysis. In the central laboratory facility the office areas need a slightly positive air pressure to prevent the seepage of toxic fumes from the analytical area into other locations within the laboratory. The primary activity related to laboratory safety should be to promote a general awareness of the need for considering safety on the part of all laboratory workers.

SELECTED REFERENCES

Briggs, R., Page, H.R.S. and Schofield, J.W., "Improvements in Sensor and System Technology", International Workshop on Instrumentation and Control for Water and Wastewater Treatment and Transport Systems, 1977, Water Research Centre, Stevenage Lab, London, England.

Brinkhoff, H.C., "Continuous On-Stream Monitoring of Water Quality", International Workshop on Instrumentation and Control for Water and Wastewater Treatment and Transport Systems, 1977, Philips Science and Industry Division, London, England.

Brown, E., Skougstad, M.W. and Fishman, M.J., "Techniques of Water Resources Investigations of the United States Geological Survey", Chapter A1--Methods for Collection and Analysis of Water Samples for Dissolved Minerals and Gases, Book 5--Laboratory Analysis, 1970, 160 pp., U.S. Geological Survey, Washington, D.C.

Cavagnaro, D.M., "Automatic Acquisition of Water Quality Data. Volume 1. 1970-1975 (A Bibliography with Abstracts)", NTIS/PS-76/0670, 1977, National Technical Information Service, U.S. Department of Commerce, Springfield, Virginia.

Cavagnaro, D.M., "Automatic Acquisition of Water Quality Data. Volume 2. 1976-July, 1978 (A Bibliography with Abstracts)", NTIS/PS-78/0887/6, 1978, National Technical Information Service, U.S. Department of Commerce, Springfield, Virginia.

Cavagnaro, D.M., "Automatic Acquisition of Water Quality Data. Volume 2. 1976-1977 (A Bibliography with Abstracts)", NTIS/PS-79/1054/0, 1979, National Technical Information Service, U.S. Department of Commerce, Springfield, Virginia.

Cavagnaro, D.M., "Automatic Acquisition of Water Quality Data. Volume 3. 1978-August, 1979 (A Bibliography with Abstracts)", NTIS/PS-79/1055/7, 1979, National Technical Information Service, U.S. Department of Commerce, Springfield, Virginia.

Cavagnaro, D.M. and Hundemann, A.S., "Automatic Acquisition of Water Quality Data. 1978-August, 1980 (Citations from the NTIS Data Base)", PB80-815772, 1980, National Technical Information Service, U.S. Department of Commerce, Springfield, Virginia.

Davis, P.E., "A Study of the Reliability of Continuous Water Quality Monitoring", PB-228 872/8, 1973, National Technical Information Service, U.S. Department of Commerce, Springfield, Virginia.

Eagleson, K.W., Morgan, E.L. and McCollough, N., "Water Quality Monitoring Using an Automated Biomonitor and NASA's GOES Satellite", Journal of Tennessee Academy of Science, Vol. 53, No. 2, Apr. 1978, p. 76.

Fisher, P.D. and Siebert, J.E., "Integrated Automatic Water Sample Collection System", Journal of the Environmental Engineering Division, American Society of Civil Engineers, Vol. 103, No. EE4, Aug. 1977, pp. 725-728.

Grana, D.C. and Haynes, D.P., "Remote Water Monitoring System", PATENT-4 089 209, 1977, U.S. Patent Office, Washington, D.C.

IHD-WHO Working Group on the Quality of Water, "Water Quality Surveys", Studies and Reports in Hydrology - 23, 1978, 350 pp., United Nations Educational, Scientific and Cultural Organization, Paris, France, and World Health Organization, Geneva, Switzerland.

Jackson, H.W., "Biological Field Methods", in Training Course Manual entitled Aquatic Biology for Engineers, 1963b, U.S. Public Health Service, Cincinnati, Ohio.

Kittrell, F.W., "A Practical Guide to Water Quality Studies of Streams", CWR-5, 1969, 135 pp., Federal Water Pollution Control Administration, U.S. Department of Interior, Washington, D.C.

Koehler, F.A., "Simple Sampler Activation and Recording System", Journal of the Environmental Engineering Division, American Society of Civil Engineers, Vol. 104, No. EES, Oct. 1978.

Lillesand, T.M., Scarpace, F.L. and Clapp, J.L., "Photographic Quantification of Water Quality in Mixing Zones", NASA-CR-137268, 1973, University of Wisconsin, Madison, Wisconsin.

National Technical Information Service, "Automatic Acquisition of Water Quality Data. 1978-January, 1982 (Citations from the NTIS Data Base)", PB82-804790, 1982, U.S. Department of Commerce, Springfield, Virginia.

Rainwater, F.H. and Thatcher, L.L., "Methods for Collection and Analysis of Water Samples", Water-Supply Paper 1454, 1960, 301 pp., U.S. Geological Survey, Washington, D.C.

Rogers, R.H., "Application of Landsat to the Surveillance and Control of Lake Eutrophication in the Great Lakes Basin", NASA-CR-143409, 1975, National Aeronautics and Space Administration, Washington, D.C.

Thiruvengadachari, N.G. et al., "Some Ground Truth Considerations in Inland Water Surveys", International Journal of Remote Sensing, Vol. 4, No. 3, 1983, pp. 537-544.

Wallace, J.W., Lovelady, R.W. and Ferguson, R.L., "Design, Development, and Field Demonstration of a Remotely Deployable Water Quality Monitoring System", EPA-600/4-81-061, 1981, U.S. Environmental Protection Agency, Hampton, Virginia.

Welby, C.W., "Use of Multispectral Photography in Water Resources Planning and Management in North Carolina", UNC-WRRJ-76-115, 1976, University of North Carolina, Chapel Hill, North Carolina.

CHAPTER 7

DATA ANALYSIS AND PRESENTATION

Another important element in a river monitoring program is a system for data recording and storage as well as considerations related to data analysis and presentation. Analytical data obtained in water quality surveys should be filed in an orderly and systematic manner to ensure rapid access to the data when it is required by study managers, research workers, and other potential users. The data filing system should be organized in a manner which allows for the easy assemblage of information for cursory review as well as more detailed analysis. Usage of the data and its analysis and interpretation is also necessary as a part of report preparation. This chapter is focused on data storage and processing systems, brief information on data interpretation, water quality indices, and the preparation of reports on water quality studies.

DATA STORAGE AND PROCESSING SYSTEMS

Water quality data may be stored manually on laboratory cards or on chemical analyses report forms and filed in standard filing cabinets, or the data may be stored automatically on magnetic tapes through the use of computer equipment (IHD-WHO Working Group on the Quality of Water, 1978). In the initial stages of a monitoring program, the volume of data is likely to be small, and it would be more advisable to begin with a simple manual data storage system. In the longer term, it may be desirable to consider the usage of automatic systems for data handling.

DATA INTERPRETATION

Once water quality data have been collected and assembled in data storage systems, the next step is to interpret the data with respect to study objectives, specific questions, environmental problems, and water resource management requirements. To serve as an example of some of the types of questions that might be raised related to water quality data, the following list is provided (IHD-WHO Working Group on the Quality of Water, 1978):

(1)   What is the water quality at any specific location or area?

(2)   What are the water quality trends in the region; is the quality improving or getting worse?

(3)   How do certain parameters relate with one another at given sites, for example, how does specific conductance relate with total dissolved solids, and how do these relate with stream discharge?

(4)   Are sampling frequencies adequate and are sampling stations suitably located to represent water quality conditions in an area?

(5)   What are the total mass loadings of materials moving in and out of water systems and from what sources do these originate?

Although general water quality conditions may be ascertained by scanning data presented in a tabulated form, the most reliable method is to conduct statistical analyses for purposes of data summary (IHD-WHO Working Group on the Quality of Water, 1978). A complete discussion of statistical tests is considered to be beyond the scope of this book; however, a good presentation of statistical tests along with example calculations is available (IHD-WHO Working Group on the Quality of Water, 1978). Some examples of some statistical tests which could be used in analyzing water quality data and subsequent interpretation are mean values, standard deviation, cumulative frequency distribution, mass loadings, least square fitting to straight lines or curves, comparison of average values, and comparison of cumulative distributions. Both tabular as well as graphical displays can be an aid in data presentation and interpretation.

WATER QUALITY INDICES

Due to the potentially large number of water quality parameters to be monitored in a river study, presentation of summary information related to these parameters will be difficult. One approach which should be considered is the use of an empirical index which combines the data from several parameters into one common numeric indicator of water quality. This numeric indicator represents an attempt to consider the relative importance of each of the water quality parameters as well as their characteristics at specific locations. The usage of water quality indices has been increasing in the United States as well as in other countries.

A survey of the extent to which water quality indices are being used in the United States has been recently conducted (Ott, 1978a). The study approach consisted of a review of the indices published in the literature; a survey of 51 state agencies, 9 interstate commissions, and 10 U.S. Environmental Protection Agency regional offices; and brief case studies of agencies that have developed their own water quality indices. Over 20 physical and chemical water quality indices have been published in various journals, symposia proceedings, and technical reports. It is possible to classify these indices into four general groups: (1) indices of general water quality, (2) indices for specific water uses, (3) indices for planning, and (4) statistical approaches. The published indices differ greatly in the numbers and types of variables included, mathematical structures, scales (whether the numbers increase or decrease with pollution), and overall index ranges.

Of the 51 state agencies surveyed (including the District of Columbia), 10 are classified as index users. This means that the agency has used a water quality index in an official publication or has applied it in a large scale study extending over a year or more. Two interstate commissions are also index users. Of the agencies using water quality indices, the most frequently used is the National Sanitation Foundation Water Quality Index (WQI). An additional 17 states and 1 interstate commission were found to be considering the use of water quality indices, with 14 indicating that they were giving consideration to using the WQI. The three most frequently mentioned purposes for using water quality indices by the users were: (1) preparation of annual reports, (2) public information presentations, and (3) analysis of water quality trends.

The WQI was developed in 1970 using a formal procedure based on the Delphi technique (Ott, 1978b). The Delphi technique consisted of combining the opinions of a large panel of water experts from throughout the United States by polling them by mail through the use of questionnaires. The results from these questionnaires were tabulated and reported to each member, enabling them to see how their own responses compared with those of the group as a whole. Then each member was polled again to arrive at a final consensus. The opinion research technique consisted of the following elements (Engineering Division, 1976):

(1) A panel of 142 experts was selected from various people closely associated with the water quality field, and each expert was sent a series of questionnaires.

(2) The first questionnaire invited them to select significant parameters of water pollution from a list of 35, and also to rate their significance on a scale of importance. Opportunity was given to include additional parameters.

(3) The second questionnaire incorporated the results of the first and asked each participant if, after comparing his responses to the general trend, he wished to revise his opinion. The participants were also asked to stipulate the 15 "most important" parameters for inclusion in a water quality index.

(4) The third questionnaire presented the parameters which had emerged from the previous work as being the most significant, and it was requested that curves assigning variations in water quality for different parametric values should be drawn for each parameter. They were also asked to alot a relative significance grading on a five point scale to each parameter. This enabled weighting for the parameters to be obtained.

(5) Finally, a set of 9 "average" parameter-quality curves with 80 percent confidence limits was compiled from the returns. This set of curves are shown in Figures 21 through 29 (Ott, 1978b). The average is the dark line and the 80 percent confidence limits are represented by the shaded area. Although some of the curves showed a remarkably low variation of opinion, others, for example fecal coliforms, displayed wide variations. The curves shown in Figures 21 through 29 can be called functional curves.

One of the additional elements related to the index is the assignment of relative importance weights to the nine final selected parameters. The following procedure was used to achieve the importance weighting. The arithmetic means of the significance ratings were calculated for all variables rated in the third questionnaire. The mean of these ratings is shown in Table 26 (Ott, 1978b). Temporary weights then were derived by dividing the significance rating of each variable into the rating for the variable with the highest significance rating (the rating used on questionnaire 3 was on a scale from 1 representing the highest relative significance to 5 representing the lowest relative significance). Finally, each temporary weight was divided by the sum of the temporary weights, giving the sub-index weights shown in the last column of Table 26.

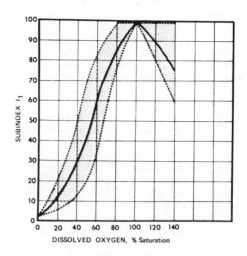

Figure 21: Functional Relationship for Dissolved Oxygen (Ott, 1978b)

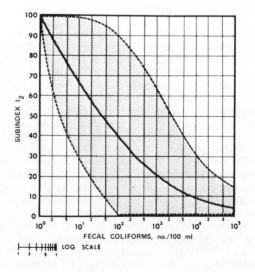

Figure 22: Functional Relationship for Fecal Coliforms (Ott, 1978b)

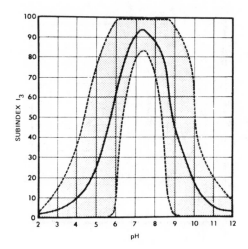

Figure 23: Functional Relationship for pH (Ott, 1978b)

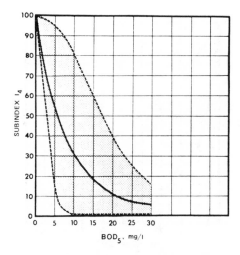

Figure 24: Functional Relationship for BOD (Ott, 1978b)

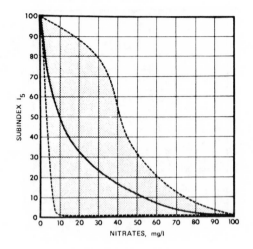

Figure 25: Functional Relationship for Nitrates (Ott, 1978b)

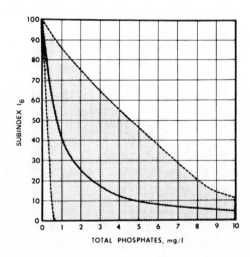

Figure 26: Functional Relationship for Total Phosphates (Ott, 1978b)

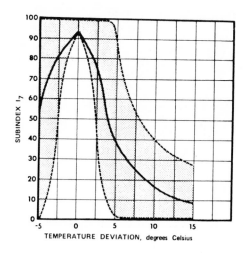

Figure 27:   Functional Relationship for Temperature Deviation from
             Equilibrium (Ott, 1978b)

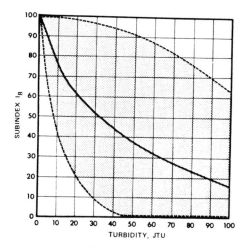

Figure 28:   Functional Relationship for Turbidity (Jackson Turbidity
             Units) (Ott, 1978b)

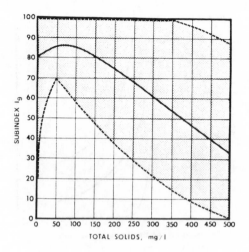

Figure 29: Functional Relationship for Total Solids (Ott, 1978b)

Table 26: Significance Ratings and Weights for Nine Pollutant Variables
          WQI (Ott, 1978b)

Variable	Mean of all Significance Ratings Returned by Respondents	Temporary Weights	Final Weights
Dissolved Oxygen	1.4	1.0	0.17
Fecal Coliforms	1.5	0.9	0.15
pH	2.1	0.7	0.12
5-Day Biochemical Oxygen Demand	2.3	0.6	0.10
Nitrates	2.4	0.6	0.10
Phosphates	2.4	0.6	0.10
Temperature	2.4	0.6	0.10
Turbidity	2.9	0.5	0.08
Total Solids	3.2	0.4	0.08
Total			1.00

Two formulations have been considered for the WQI by the National
Sanitation Foundation. One represents an arithmetic weighted approach and
the other a geometric weighted approach. The arithmetic weighted approach is
shown by the following formulation:

$$WQI_a = \sum_{i=1}^{n} I_i W_i$$

while the geometric approach is as follows:

$$WQI_g = \prod_{i=1}^{n} I_i W_i$$

The arithmetic weighted approach involves summing the individual products of the water quality rating and the corresponding weighting. The geometric weighted approach is determined by multiplying together each water quality rating raised to the power of its weighting. Table 27 provides an example of calculations for a water quality index using both approaches. When the arithmetic weighted approach is used, the resultant water quality index would be significantly changed if only one variable shows an extremely poor water quality. The geometric weighted approach tends to reduce this effect. Subsequent studies as to which is the best formulation, again using a panel of experts, indicated that the geometric weighted approach agreed better with expert opinion than did the arithmetic weighted approach. Either approach could be used as long as the users were clear as to its limitations.

Table 27: Calculations for Water Quality Index

Variable	Measurement	$I_i(a)$	$W_i(b)$	$I_i W_i$	$I_i^{W_i}$
DO	60%	60	0.17	10.2	2.01
Fecal Coliforms	10^3	20	0.15	3.0	1.57
pH	7	90	0.12	10.8	1.72
BOD_5	10	30	0.10	3.0	1.41
NO_3	10	50	0.10	5.0	1.48
PO_4	5	10	0.10	1.0	1.26
Temperature Deviation	5	40	0.10	4.0	1.45
JTU	40	44	0.08	3.5	1.35
Total Solids	300	60	0.08	4.8	1.39

$$WQI_a = 45.3 \qquad WQI_g = 38.8$$

(a) Subindex values are from Figures 21 through 29.

(b) Weights are from Table 26.

One final element of the WQI is related to an ultimate classification of quality according to the numerical range. Table 28 summarizes a suggested classification based on five ranges (Ott, 1978b).

Table 28: Stream Classification Based on WQI_g (Ott, 1978b)

Numerical Range	Classification
0 – 25	Very bad
26 – 50	Bad
51 – 70	Medium
71 – 90	Good
91 – 100	Excellent

The WQI could be used in a river study, or a specific index appropriate for the area could be developed. If a specific water quality index is developed, the same general approach used in developing the National Sanitation Foundation WQI could be applied. This same approach has been used by others in the development of water quality indices, with one example being a water quality index developed in Scotland (Engineering Division, 1976). Table 29 lists some references related to water quality and biological indices, with their abstracts contained in Appendix J.

Table 29: References on Water Quality and Biological Indices

Authors (Year)	Comments
Ball and Church (1980)	Usage and limitations of water quality indices.
Booth, Carubia and Lutz (1976)	Methodology for comparative evaluation of indices.
Dunnette (1979)	Index based on geographical characteristics of river basins.
House and Ellis (1981)	Advantages of water quality indices.
Inhaber (1976)	Book describing economic, air quality, water, land, biological, aesthetic, and other environmental indices.
Ott (1978)	Book describing the structure of environmental indices, including air pollution and water pollution indices.
Polivannaya and Sergeyeva (1978)	Use of zooplankters as bioindicators of water quality.

Table 29: (continued)

Authors (Year)	Comments
Yu and Fogel (1978)	Index based on use-oriented benefits and treatment costs analysis.

REPORT OUTLINE

The end product of a river study should be one or more reports which address study findings and provide appropriate interpretation.   While this represents a work element in the future, careful planning from the on-set relative to the ultimate report can be of value, particularly as related to data organization and analysis. The report (or series of reports) should be prepared as soon as possible after completion of the field work while the details are still fresh to the sampling team as well as analytical laboratory personnel and study managers. There is no standardized outline for a study report. The specific outline will need to be developed as the study proceeds. A potential outline for a report is contained in Table 30 (IHD-WHO Working Group on the Quality of Water, 1978). This outline is not intended to be comprehensive; however, it does represent a list of topics which should be addressed as a part of the report.

Table 30:   Report Outline for Water Quality Study (IHD-WHO Working Group on the Quality of Water, 1978)

INTRODUCTION

    (1)   Objectives and terms of reference of the executing agency.

    (2)   Description of stream and its basin; its present and prospective uses and sources of pollution.

METHODOLOGY

    (1)   Description of sampling stations, outfalls and sampling procedure.

    (2)   Laboratory facilities and analytical techniques.

    (3)   Hydrological measurements.

OBSERVATIONS

    (1)   Sources of pollution.

    (2)   Stream water quality, sediment quality, and biological characteristics.

    (3)   Stream discharge and other hydromorphological measurements.

Table 30: (continued)

EVALUATION OF RESULTS

    (1)   Analysis of data.

    (2)   Formulation and evaluation of models describing the behavior of pollutants.

RECOMMENDATIONS

    (1)   Necessity of pollution abatement measures.

    (2)   Future surveillance program.

GENERAL REPORT WRITING CONSIDERATIONS

The preparation of an effective report on a water quality study requires that the writers be familiar with the study, that ample visual aids are utilized in the report, that the report is written in a fashion which communicates best with the target audience for the report (this involves consideration of whether the report should be written for technical versus nontechnical audiences), and that there is an internal review of the report prior to its publication. The principal writer or writers of the report should be familiar with the details of the study, and preferably had an active role in the field work itself (Kittrell, 1969). The writer should be thoroughly familiar with the study river, the location of sampling stations, waste sources and water uses, and the details of the sampling and analytical programs. This broad knowledge is essential for the most effective interpretation of data and report preparation.

The report should contain ample visual aids such as graphs, photographs, and data presentations in tabular form. A basic map is essential in any report of a stream study. The map should show the stream being studied and the major tributaries involved. Insignificant streams, highways, and towns that are not involved, as well as other details such as contour lines, should be omitted from the basic map since they add a great deal of detail but very little effectiveness in information communication. On the other hand, any geographical feature of the study area that is mentioned in the report should be shown on the basic map. Photographs can be a useful tool in furnishing visual impressions of the features of a stream. Photographs of the sampling equipment and sample collection operations can also be effective. If there are locations in the study river characterized by evidences of pollution, photographs of floating materials and other unsightly conditions can also be an effective tool in information communication.

Since a large amount of information will be collected from most river studies, there will be the need for extensive data presentations in tabular form in the report. A list of sampling stations, with descriptions, and other pertinent features such as flow gaging stations, tributary stream confluences, and points of water usage or wastewater discharge can be presented in a tabular form. Data on wastewater discharges, including the name of the town or industry, the location, the flow, and other characteristics can also be effectively presented in a tabular format.   Analytical data on various physical, chemical, and bacteriological characteristics of the river waters can also be presented and summarized in tabular forms. It might be appropriate to summarize the data by average, maximum, and minimum values along with more complete tabular presentations contained in appendices.

The use of graphic presentations of the study findings can also be an effective tool for information presentation and interpretation. Specific charts or figures which would be appropriate for the study river will have to be developed as the study progresses, but it is worthwhile to begin to consider the types of graphical presentations which would be effective. In conjunction with the use of visual aid materials, it should be noted that the biological characteristics of the study river should be incorporated along with traditional water quality characteristics. In many cases biological findings are reported separately as though they were adjunct to the other data. Separate biological sections have been placed in appendices, and an integrated analysis and interpretation relative to water quality is sometimes not performed. Since the biological characteristics of the study river are an integral element of the overall ecosystem, integration of this information with water quality and sediment characteristics is vital. Biological findings can also be presented in an effective manner through the use of visual aid materials.

An important issue which must be identified early in the preparation of a report is the target audience for the report. The information presentation and summary should differ if the target audience is the general public as opposed to more technically oriented audiences represented by the governmental agencies, private industries, and others. The report should avoid the use of technical language, and details as much as possible, if it is designed for a nontechnical audience. In order to achieve this, it may be necessary to include a glossary of terms, even if the report is to be reviewed by technical audiences. The purpose of the report is to communicate information in the most effective manner, and not simply to confuse audiences with in-depth scientific knowledge in a particular area. Accordingly, the biological information and species data should be reported very carefully so as to be able to be understood by nonbiologists.

The text of the report should immediately focus on the key objectives and findings of the study. Lengthy descriptions of methods of laboratory analysis, sampling, flow measurement, and calculations can be more appropriately included in appendices.

An effective approach which can be used to ensure that the report communicates in an appropriate manner is to subject the report to internal review prior to its publication. This internal review should be done by both technical as well as nontechnical individuals, with at least one person reviewing the report who is not familiar with the study or study area. If the report effectively communicates to the person who is unfamiliar with the study, then this represents a good measure of report effectiveness (Kittrell, 1969).

SELECTED REFERENCES

Ball, R.O. and Church, R.L., "Water Quality Indexing and Scoring", Journal of the Environmental Engineering Division, American Society of Civil Engineers, Vol. 106, No. EE4, Aug. 1980, pp. 757-771.

Booth, W.E., Carubia, P.C. and Lutz, F.C., "A Methodology for Comparative Evaluation of Water Quality Indices", 1976, Worcester Polytechnic Institute, Worcester, Massachusetts.

Dunnette, D.A., "A Geographically Variable Water Quality Index Used in Oregon", Water Pollution Control Federation Journal, Vol. 51, No. 1, Jan. 1979, pp. 53-61.

Engineering Division, "Development of a Water Quality Index", ARD 3, Dec. 1976, 62 pp., Scottish Development Department, Edinburgh, Scotland.

House, M. and Ellis, J.B., "Water Quality Indices: An Additional Management Tool", Water Science and Technology, Vol. 13, No. 7, 1981, pp. 413-423.

IHD-WHO Working Group on the Quality of Water, "Water Quality Surveys", Studies and Reports in Hydrology - 23, 1978, 350 pp., United Nations Educational, Scientific and Cultural Organization, Paris, France, and World Health Organization, Geneva, Switzerland.

Inhaber, H., Environmental Indices, John Wiley and Sons, Inc., 1976, New York, New York.

Kittrell, F.W., "A Practical Guide to Water Quality Studies of Streams", CWR-5, 1969, 135 pp., Federal Water Pollution Control Administration, U.S. Department of Interior, Washington, D.C.

Ott, W.R., "Water Quality Indices: A Survey of Indices Used in the United States", EPA-600/4-78-005, 1978a, 128 pp., U.S. Environmental Protection Agency, Washington, D.C.

Ott, W.R., Environmental Indices--Theory and Practice, 1978b, Ann Arbor Science Publishers, Inc., Ann Arbor, Michigan, pp. 202-213.

Ott, W.R., Environmental Indices--Theory and Practice, 1978, Ann Arbor Science Publishers, Inc., Ann Arbor, Michigan, 1978.

Polivannaya, M.F. and Sergeyeva, O.A., "Zooplankters as Bioindicators of Water Quality", Hydrobiological Journal, Vol. 14, No. 3, 1978, pp. 39-43.

Yu, J.K. and Fogel, M.M., "The Development of a Combined Water Quality Index", Water Resources Bulletin, Vol. 14, No. 5, Oct. 1978, pp. 1239-1250.

APPENDIX A

SUMMARY OF NETWORK DESIGN PROCEDURES
(ABSTRACTED FROM SANDERS, 1980)

SUMMARY OF NETWORK DESIGN PROCEDURES

Network design will never be quantified by a set of hard and fast rules or steps; therefore, the following is intended more as a general guideline than as a set of rules for design. The more subjective aspects of network design are listed along with the more quantifiable aspects in order to give a broader perspective of network design.

STEP 1. DETERMINE MONITORING OBJECTIVES AND RELATIVE IMPORTANCE OF EACH

The reasons for monitoring water quality should be determined as precisely as possible. This may require the network designer to formulate the objectives in his own words prior to approaching the ultimate users of the data and/or information. Such formulations provide a specific basis for discussion.

If the network tries to meet several objectives, each should be determined and prioritized relative to the others. Conflicting objectives require the compromise of many design specifications later in the design process, and a relative weighting of objectives provides the designer with guidance in compromising. Once the specific objectives of a water quality monitoring network have been articulated, the designer must then translate these objectives into specific data requirements which can be fulfilled by the proposed network design.

STEP 2. EXPRESS OBJECTIVES IN STATISTICAL TERMS

Translating objectives from words to statistics at the first of the design process permits the users of data and information to specify the accuracy they need in quantifiable terms while at the same time providing the network designer with a more objective basis for future design calculations. If one of the monitoring activities is associated with enforcement or other statutory needs, it is very important that the design of the network and the resultant data obtained are compatible with existing and future legislation.

Water quality monitoring is a statistical sampling process and thus there will be uncertainty associated with the final results. The acknowledgment of this fact will greatly assist network designers in developing meaningful and useful designs.

STEP 3. DETERMINE BUDGET AVAILABLE FOR MONITORING AND AMOUNT TO BE ALLOCATED FOR EACH OBJECTIVE

Realistic objectives cannot be formulated without acknowledging the limits within which one must work. Economic limits on the number of samples, number of stations, etc., will greatly influence the ability of a network to reach its statistically defined objectives. Recognizing the economic limits early will, again, permit the data users to participate in deciding where the compromises will be made. One compromise may be to trade off percent areal coverage (a large number of stations) for more intense sampling at a small number of stations.

If the network is multi-objective, an attempt should be made to determine the amount of the budget to be allocated to each objective--the ultimate measure of the importance a data or information user places on the objective.

STEP 4. DEFINE THE CHARACTERISTICS OF THE AREA IN WHICH THE MONITORING IS TO TAKE PLACE

The geographical and hydrological characteristics of an area to be monitored need to be well defined prior to initiation of network design calculations. Natural salt springs, industrial concentrations, population centers, flow patterns, irrigation schedules, etc. will all influence the network design process. The extent of influence will depend upon the objectives of the network.

Networks which must cover a number of watersheds, or portions thereof, require careful consideration of the geographical and hydrological differences and how these differences may preclude the use of one uniformly designed network for the entire jurisdiction. Instead, the sub-basin objectives and budget allocations must be determined and each sub-basin then treated as its own separate network--a sub-network of the overall network.

STEP 5. DETERMINE WATER QUALITY VARIABLES TO BE MONITORED

Variables measured by a water quality monitoring network are highly dependent upon the objectives, basin characteristics and economics of the network. For example, a regulatory water quality monitoring network would perhaps be mainly interested in those variables stated in the stream standards.

The design of a network must center around water quality; however, water quality can be defined in terms of one variable or 1000 variables. Design of a water quality monitoring network, unless only one variable is considered, must account for the different statistical behavior of the different variables. This will require compromising of some form among the variables.

Both sampling location and sampling frequency can be developed independently of frequencies for analyzing samples in the laboratory. Both location and frequency are specified for the collection of the water sample--the analyses are made later. However, both criteria are affected by the behavior of the water quality variable being monitored. For example, sampling once a week at a single point in a river may be more than adequate for monitoring a relatively stable river temperature, but may be hardly adequate for monitoring rapidly varying coliform bacteria concentrations. Therefore, before a water quality monitoring network can be designed in a systematic fashion, the variables to be monitored should be specified so that their natural and/or man-made variation in time and space can be considered when designing the monitoring network. In addition to considering the water quality variables of interest, their respective units should be delineated as well. The network design varies tremendously if a daily mean (flow weighted) concentration is needed rather than an instantaneous grab sample concentration--the former being a result of several samples with flow measurements equally spaced during a 24-hour period, while the latter is only a single sample (generally in the daytime between 8:00 a.m. - 4:30 p.m.)

STEP 6. DETERMINE SAMPLING STATION LOCATIONS

The location of a permanent sampling station in a water quality monitoring network is probably the most critical aspect of network design, but it is all too often never properly addressed. Expediency and cost compromises lead in many cases to sampling from bridges or near existing river gaging stations. Whether the single grab sample from the bridge or the gaging station is truly representative of the water mass being sampled is not known, but it generally is assumed to be representative by both the collectors and users of the water quality data. Using river stage for estimating discharge, measurement anywhere in the lateral transect would indicate the exact river discharge. However, this does not necessarily follow when measuring water quality variable concentrations. In fact research indicates the opposite, that a single sample will rarely be indicative of the average water quality in a river's cross section.

Sampling locations for a permanent water quality network can be classified into two levels of design: macrolocation and microlocation, the former is a function of the specific objectives of the network while the latter is independent of the objectives but a function of the representativeness of the water sample to be collected.

The macrolocation within a river basin is usually determined by areas of major pollution loads, population centers, etc. Macrolocation can be specified, as well, according to percent areal coverage using basin centroids. This methodology locates sampling points in a systematic fashion, maximizing information for the entire basin with a few strategically located stations. Figure 30 is an example of locating sampling stations using basin centroids and sub-basin centroids with percent areal coverage as the criteria.

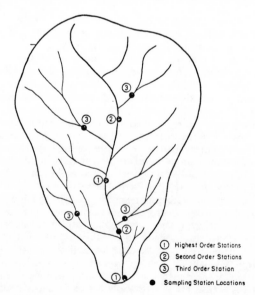

①	Highest Order Stations
②	Second Order Stations
③	Third Order Station
●	Sampling Station Locations

Figure 30:   Macrolocation of Sampling Stations within a River Basin Using the Percent Areal Coverage as the Criteria Specifying Location

The percentage of areal coverage specified by the monitoring agency is defined as the number of sampling stations divided by the magnitude of the basin. Intrinsic to this objective procedure is the concept of a sampling station hierarchy that orders the importance of each sampling station in the basin. This provides a realistic methodology in which a rational implementation program can proceed: the most important stations (highest order) are built first and as the resources become available additional stations can be built. As each succeeding hierarchy of stations is established, the percentage of areal coverage is increased.

Once the macrolocations within a river basin are established, the microlocation is then determined. The macrolocation specifies the river reach to be sampled while the microlocation specifies the point in the reach to be sampled. The latter point is the location of a zone in the river reach where complete mixing exists and only one sample is required from the lateral transect in order to obtain a representative (in space) sample. The zone of complete mixing can be estimated using various methodologies.

Assuming that a point source pollutant distribution in a stream approximates a Gaussian distribution, and that boundaries can be modeled using image theory, the following equation can predict the distance downstream in a straight, uniform channel from a point source pollutant to a zone of complete mixing:

$$L_y = \frac{\sigma_y{}^2 u}{2 D_y}$$

where L_y = mixing distance for complete lateral mixing,

σ_y = distance from point source to farthest lateral boundary,

u = mean stream velocity,

D_y = lateral turbulent diffusion coefficient.

Unfortunately, there may not always be points of complete mixing due in part to the random nature of the aforementioned mixing distance, inapplicability of the assumptions used in the determination of the mixing distance, or, more often than not, not enough river length or turbulence to assure complete mixing within the specified river reach.*

If there is not a completely mixed zone in the river reach to be sampled, there are three alternatives: (1) sample anyway at a single point and assume it is representative (this is the general procedure being applied today); (2) do not sample the river reach at all, because the data which would be obtained does not represent the existing river quality, but only the quality of the sample volume collected--in other words, the data is useless; (3) sample at several points in the lateral transect collecting a composite mean, which would indeed be representative of the water quality in the river at that point in time and space.

---

*It should be noted that field verification of a completely mixed zone prior to locating a permanent sampling station can be easily done by collecting multiple samples in the cross section and analyzing the data using the well known one- or two-way analysis of variance techniques.

$t_{\alpha}/2$ = constant which is a function of the level of significance and the number of samples,

S = standard deviation of the water quality concentrations,

R = specified half-width of the confidence interval of the annual mean.

Using the same assumption, that the water quality variable is iid, the number of samples per year can be specified as a function of other data analysis procedures as well. For example, if annual means were to be tested for significant differences between years, the number of samples needed to detect a given level of change can be specified.

A much more sophisticated procedure representing a higher level of statistical analysis is to recognize that water quality variables may not be independent, but highly dependent, not identically distributed, but seasonally variable, and to determine sampling frequency as a function of the variability of the water quality time series after trend and periodic components have been removed. Unfortunately, other than mean daily discharge, temperature and conductivity, data bases of water quality variables of sufficient number, reliability and length are not usually available for application of this procedure.

Once a consistent sampling frequency criterion is selected, it can be utilized to objectively distribute sampling frequencies within a water quality monitoring network. For example, the expected half-width of the confidence interval about the annual mean approach can be applied basin-wide (for specifying sampling frequencies) in a consistent fashion by specifying equality of these expected half-widths at each sampling station. Thus, stations where water quality varies tremendously will be sampled more frequently than stations where the water quality varies little. With reference to Figure 31, which is a plot of the expected half-width of the confidence interval of mean log river flow versus the number of samples per year, the number of samples collected at each station within the river basin for a given R is determined by drawing a horizontal line through R and reading the number of samples on the abscissa axis below the intersections of the horizontal line with each curve. Figure 31 may also be used in an iterative fashion to specify sampling frequencies at each station when a total number of samples from the basin is specified. For example, if only n samples per year need to be collected and analyzed, a value of R is assumed and a line is drawn horizontally; the number of samples specified by the intersection of the horizontal line with the curves are summed and compared to n. If the sum were not equal to n then another estimate of R would be made until the sum of all the samples were equal to n.

The expected half-width of the confidence interval of the annual mean is not the only statistic that can be used to specify sampling frequencies; the expected half-width divided by the annual mean is a measure of relative error and may be more appropriate when assigning sampling frequencies in a network where water quality varies tremendously from river to river.

When developing sampling frequencies, one must keep in mind two very important cycles which can have immense impact on water quality concentrations--the diurnal cycle and the weekly cycle. The effect of the diurnal cycle (which is a function of the rotation of the earth) can be eliminated by sampling in equal time intervals for a 24-hour period and the effect of the weekly cycle (which is a function of man's activity) can be eliminated by

If the sample were not representative of the water mass, the frequency of sampling as well as the mode of data analysis, interpretation and presentation and the realistic use of the data for objective decision making would be inconsequential. In spite of this fact, criteria to establish station locations for representative sampling has received relatively little attention from state and federal agencies responsible for water quality monitoring.

STEP 7. DETERMINE SAMPLING FREQUENCY

Once sampling stations have been located so that samples collected are representative in space, sampling frequency should be specified so that the samples are representative in time.

Sampling frequency at each permanent sampling station within a river basin is a very important consideration in the design of a water quality monitoring network. A large portion of the costs of operating a monitoring network is directly related to the frequency of sampling. However, the reliability and utility of water quality data derived from a monitoring network is likewise related to the frequency of sampling. Significant as sampling frequency is to detecting stream standards violation, maintaining effluent standards, and estimating temporal changes in ambient water quality, very little quantitative criteria designating appropriate sampling frequencies have been applied to the design of water quality monitoring networks. In many cases, professional judgment and cost constraints provide the basis for sampling frequencies. All too often, frequencies are the same at each station and based upon routine capabilities, once-a-month, once-a-week, etc. At present, these possibly may be the only practical means to implement a sampling program, considering the statistical background of data collectors. There do, however, exist many quantitative, statistically meaningful procedures to specify sampling frequencies at each station. The methods include specifying frequencies as functions of the cyclic variations of the water quality variable (Nyquist frequency), the drainage basin area and the ratio of maximum to minimum flow, the confidence interval of the annual mean, the number of data per year for testing hypotheses, and the power of a test measuring water quality intervention.

All of the aforementioned procedures can be applied to the design of a water quality monitoring network with each requiring a different level of statistical sophistication in terms of data requirements, necessary assumptions and limitations, and statistical expertise of the users.

One of the simplest approaches to determine sampling frequency is to assume that the water quality variable concentrations are random, independent and identically distributed (iid) and determine the number of samples per year as a function of an allowable (specified) confidence interval of the mean annual concentration (this is analogous to the procedure for determining how many analyses of a water sample should be made to determine a reasonable estimate of the mean water quality variable concentration).

$$n = \left[\frac{t_{\alpha/2} S}{R} \right]^2$$

where n = number of equally spaced samples collected per year,

specifying that sampling intervals for a network cannot be multiples of seven--occasional sampling on weekends would be necessary.

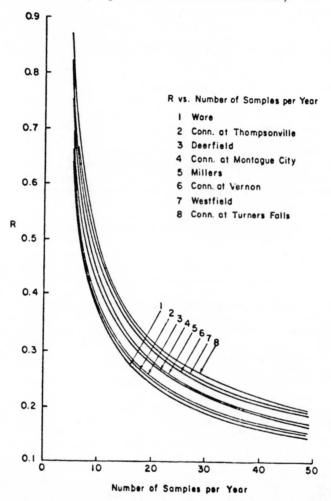

Figure 31: A Plot of Number of Samples per Year of the Expected Half-width of the Confidence Interval of Mean Log Flow, R, versus Number of Samples for Several Rivers in the Connecticut River Basin

Perhaps, the major impact of network design (in terms of variables to be monitored, sampling location, sampling frequency, and the operational monitoring functions) is in the area of data analysis and, consequently, ultimate value of the monitoring network information. Any sampling program that is to generate conclusive results from observing a stochastic time series (water quality concentrations) must be well planned and statistically designed. Statistically designed implies that the sampling is planned (in proper locations and numbers) so that the statistical analysis techniques chosen will be able to yield quantitative information. Thus, the data analysis techniques (level and type of statistics) to be used must be defined in order to know how to compute proper sampling frequencies, locations, etc.

STEP 8. COMPROMISE PREVIOUS OBJECTIVE DESIGN RESULTS WITH SUBJECTIVE CONSIDERATIONS

Station locations, sampling points, and sampling frequencies objectively computed in previous steps must now be evaluated in light of the access to the water, costs of acquiring the sample and rounding off frequencies to match practical sampling routes. Such compromising must be minimized, but to say that it should not be done is not realistic. Again, trade-offs must be considered--having a crew sample on weekends must be balanced against the problem of nonrepresentative data, and sampling from a bridge balanced against the cost of obtaining access elsewhere. Such compromises should be recorded and attached to any reporting of data from the network.

STEP 9. DEVELOP OPERATING PLANS AND PROCEDURES TO IMPLEMENT THE NETWORK DESIGN

The actual day-to-day operation of a network includes many functions of which network design is a predecessor. These operational and informational functions, to be performed smoothly over time and changing personnel, must be well defined and documented. This requires the network designer to develop sample collection routes and schedules, sample collection procedures and forms, laboratory analysis recording procedures and forms, data handling procedures and forms, etc. The network design should be documented well enough so that the monitoring operation can easily function in a uniform manner regardless of the personnel involved since they will surely change over time.

STEP 10. DEVELOP DATA AND INFORMATION REPORTING FORMATS AND PROCEDURES

This step is an extension of Step 9; however, its importance to the success of any network design cannot be overemphasized. The report formats and distribution procedures should be closely identified with the network's objectives, and, as a result, should be defined in consultation with the users of the results from the network.

Lack of reporting formats and procedures is a common problem with many networks today and is a sure sign that data collection may have become an end unto itself. Regular communication between gatherer and user ensures that a network's results are properly placed and meet the information needs for which the network was established.

Reporting formats will vary greatly, depending upon the network's purpose, the primary users and the budget available. For the same network there may be monthly reports of a one or two page nature, annual reports encompassing considerable detail and, therefore, length, and special reports needed to meet special information requests.

Proper planning (network design) of reporting formats and distribution procedures can greatly reduce the staff time needed to generate reports. Computer plots of data, computer printouts in a reporting format, standard reporting forms, etc. are examples of ways to expedite reporting of data and information. Where timing of information distribution is critical, such procedures are almost always required if reporting is not to disrupt normal network operations.

STEP 11. DEVELOP FEEDBACK MECHANISMS TO FINE TUNE THE NETWORK DESIGN

As part of the data and information reporting, specifically, and as part of the entire network design, in general, a means of receiving and utilizing feedback, both solicited and unsolicited, must be provided in order to "fine tune" the network's design. No water quality monitoring network design can be assumed fixed. There are always reasons for altering a network's design--changing objectives, changing technology, new data users, etc. There are some networks which may have components that change little over time (such as a network measuring long term trends), but, in general, procedures to accommodate change must be incorporated into network design.

Feedback can be incorporated into the reporting of information by spelling out in standard form the means by which a reader can report any comments he or she may have regarding the data and/or information (and, thus, the network design). Such unsolicited comment or feedback must be recorded, reported, evaluated and answered, if not to the person making the comment, to those responsible for the network's existence.

Solicited feedback can be obtained via planned questionnaires regarding the network's design. Soliciting feedback should be a regularly scheduled (e.g., every 5 years) activity associated with the network design. All aspects of the monitoring network system should be included in such an evaluation.

STEP 12. PREPARE A NETWORK DESIGN REPORT

Network designers are generally network oeprators, consequently, their design evaluations, procedures, etc. are performed and implemented without a report being prepared. Lack of network design documentation creates many problems associated with water quality monitoring today.

Whenever a network designer is designing a new network or evaluating and modifying an existing network, he should carefully describe the design process and findings in a Network Design Report. Such a report would contain the results of the previously described 11 steps in the design process. If the design of a network were contracted out, the monitoring personnel would expect such a report since the design has to be communicated between the designers and operators. Since water quality monitoring personnel are generally quite mobile, design and implementation done in-house, without documentation, often leaves those who follow with very little guidance as to network design. Documentation of network designs is very much needed and should be done any time a new network is established or an existing network is evaluated.

SELECTED REFERENCE

Sanders, T.G., editor, "Principles of Network Design for Water Quality Monitoring", July 1980, 312 pp., Colorado State University, Ft. Collins, Colorado.

APPENDIX B

GENERAL MONITORING PLANNING CONSIDERATIONS

Beach, M.I. and Beach, J.S., Jr., "Sample Acquisition the First Step in Water Quality Monitoring", International Workshop on Instrumentation and Control for Water and Wastewater Treatment and Transport Systems, 1977, N-CON Systems Co., London, England.

Wastewater sampling sites must include all pertinent substreams and exclude other streams which might dilute or interfere with analyses to be made. The site must provide a well mixed stream, preferably in an area of maximum turbulence. The site should be safe and accessible. The time period represented by the sample is important. Discrete, simple composite, and sequential composite samples may be collected either manually or automatically; flow proportional and continuous sampling are generally automated. The basic methods of sample gathering are mechanical, suction lift, and forced flow. The type of power used to operate an automatic sampler will depend on what is available at the site, convenience, and safety. The programming of sample collection may be as simple as a fixed time interval, or highly flexible. The sampler must produce a sample of adequate size for the analyses to be performed. Containers must be easy to clean and of materials that will not interfere with the analyses by leaching, absorbing, or adsorbing. Techniques for sampling wastewaters containing floating oils and solids need further development. Storm runoff research requires additional work on sampling of heavy bed loads and coarse solids. No matter how elaborate or expensive the analysis made on a sample is, it is of no value or, worse yet, misleading if the sample is not representative.

Chakrabarti, C.L. et al., "Preservation of Some Anionic Species in Natural Waters", American Water Works Association Journal, Vol. 70, No. 10, Oct. 1978, pp. 560-565.

The preservation of nitrates (NO_3-) ammonium (NH_4+), organic N, soluble inorganic P (SIP), soluble organic P (SOP), sulfide ($S-2$), cyanide ($CN-$), and fluoride ($F-$) in Rideau River, Canada, water samples is described. Currently accepted methods of preservation include storage at $4°C$, treatment with sulfuric acid (H_2SO_4), treatment with chloroform (CHC_{13}), and storage at $4°C$ after treatment adding either H_2SO_4 or CHC_{13}. The NH_4+ concentration was stabilized for at least 30 days by the addition of H_2SO_4. Sulfuric acid, however, cannot be recommended for samples containing nitrite, because of the Van Slyke reaction which produces N_2 or converts it to NO_3-. Storage at $4°C$ preserved NO_3- for as long as 20 days; the addition of CHC_{13} caused a 22 percent decrease in 7 days, although by the 30th day of storage, 16 percent of this was regained. No change occurred in the SIP and SOP concentrations for at least 30 days when samples were preserved with H_2SO_4, but significant changes took place in these values in unpreserved samples, and in those preserved by CHC_{13} or by storage at $4°C$. For $S-2$, only samples to which zinc acetate ($Zn(C_2H_3O_2)_2$) was added were stable for as long as 30 days. No change in $CN-$ concentration occurred in samples preserved by adding CHC_{13}, $Zn(C_2H_3O_2)_2$, or sodium hydroxide (pH 11) at $4°C$, or by simply storing at $4°C$. Based on this study, no single method of preservation is equally effective for the different anionic species studied.

Casey, H. and Walker, S.M., "Storage and Filtration of Water Samples", International Environment and Safety, 1981, pp. 16-17.

Ideally, the chemical composition of water being analyzed should be measured in situ. In this way the chemical speciation of the element being measured would be subject to less interference by external influences. Although there are electrochemical methods that can be used for certain parameters, for example, pH, conductivity, oxygen electrodes, and selective ion electrodes, they do not cover the complete range of analyses required. Even these measurements taken in situ can alter the chemical environment of the sampling site, for example, the oxygen electrode uses up oxygen and therefore needs to be agitated, otherwise oxygen in the water surrounding the electrode is depleted. In unpolluted waters, with the exception of anodic stripping voltametry, and neutron activation analysis, methods for the analysis of dissolved heavy metals usually require a pre-concentration step, which means in situ measurements are not possible.

Collins, W.H., "Improved Water Analysis Kit", FIRL-F3222-02/03, Dec. 1972, Franklin Institute Research Laboratories, Philadelphia, Pennsylvania.

An improved Water Test Kit which is smaller in size and simpler to use and stock than Water Quality Control Set, FSN 6630-262-7288, Standard B, has been developed. The kit has a minimum of glassware, weighs 4½ pounds, measures 9x6x6 inches, and uses techniques completely different than former methods. Eliminated is the need of preparing reagents, performing titrations and other tedious measurements. Most of the tests are conducted by dipping a paper or plastic strip into the water sample and reading the height of a column or comparing the color obtained to a chart. Water samples can be examined for pH, acidity, alkalinity, chlorine residual, chlorine demand, chloride, sulfate, turbidity and coagulation characteristics. Each kit contains enough material to examine at least 50 water samples for each of the above characteristics before requiring refill. The simplicity of the kit allows personnel with little or no training to use it effectively.

Drake, T.L., "A Systems Analysis of Water Quality Survey Design. Appendix VIII. Documentation Data Handling System Programmer's Manual", AD-A036 521, 1975, Clemson University, Clemson, South Carolina.

A Data Handling System is described which handles on-site water quality survey data measured by the U.S. Army. A sequence of programs is often executed to complete a given data handling operation. OS/8 BATCH is used to automate the specification of this sequence of operation via a BATCH input file. OS/8 BATCH provides users of the system with a batch processing monitor that is integrated into the OS/8 Monitor structure. Several system tables are used by the system software to define a particular data handling application to the system. The system definition file (SYSDEF.AR) provides information which includes lists of valid ID tags, upper and lower limits for measurement values, data formats, and various heading and column information used during report generation. The translation table file (TRNTBL.AR) contains the translation tables for the mark sense cards. The method file (METHOD.AR) specifies for each parameter the particular data conversion algorithm within the function file (FNCTN.AR) to use on input. The data files and information files are OS/8 BASIC numeric files while all other files are OS/8 ASCII files. The ID information within these numeric files are 6-bit positive integers. OS/8 BASIC user functions are provided to pack and unpack these 6-bit integers within a 36-bit floating point variable. In addition, OS/8 BASIC user functions are provided for formatted output, interization, and PDM-70 input. The chapters within this manual include a discussion of the implementation of the data handling programs, file formats, mark card design, and industry compatible type formats.

Fenlon, J.S. and Young, D.D., "Chemical Surveillance of Rivers", Water Pollution Control, Vol. 81, No. 3, 1982, pp. 343-357.

Reliable river quality information is essential to enable any river management authority to carry out its duties and it should be obtained in an economical manner. Traditionally, river water sampling has been practiced by manual methods and only in the last decade have reliable automatic monitors become available. Although they may compete for resources, they are generally complementary in the information which they provide. This paper considers the development of chemical surveillance of river water quality and assesses what its aims and objectives should be.

Heidtke, T.M. and Armstrong, J.M., "Probabilistic Sampling Model for Water Quality Management", Journal of the Water Pollution Control Federation, Vol. 51, No. 12, Dec. 1979, pp. 2916-2927.

The stochastic nature of· pollutant loadings and in-stream concentrations is incorporated into the design of a cost-effective sampling strategy for detection of stream standard violations. Sampling policies call for samples to be taken when the probability of a stream standard violation exceeds a specified threshold. Nine "fixed-time" policies are compared for effectiveness in detecting violations and for overall costs of administering the programs over a specified planning period. The optimum sampling strategy depends on the violation detection objective, i.e., the desired probability of detecting a stream standard violation. The optimum policy can also vary with the form of the damage function describing the costs of detection delays.

Landwehr, J.M., "Some Properties of the Geometric Mean and Its Use in Water Quality Standards", Water Resources Research, Vol. 14, No. 3, June 1978, pp. 467-473.

Routinely used as a water quality standard for fecal coliform measurement, the geometric mean is mathematically defined as the nth root of the product of n values. Some quality control instructions alter this formal definition by replacing values of zero counts with the value of 1 and do not precisely state how to derive the geometric mean. A statistical examination, aimed at identifying important parameters associated with the geometric mean, revealed the importance of using a standard number of samples for meaningful results. It demonstrated that the geometric mean is a function of the sample size and is very sensitive to the form and skew of the underlying distribution function. The number of samples used should be selected from the features of the distribution function, so that the expected value of the geometric mean is not overly sensitive to an increase in sample size.

May, F.C., "Vocational Education Training in Environmental Health Sciences: Collecting Stream Samples for Water Quality, Module 16", DE/OVAE-80-0088-16, 1981, Rockville, Maryland.

In this instructional module, information is provided on the collection of stream samples for specific laboratory analyses. Before beginning this module, students must know how to work safely in a laboratory; adjust the pH of a solution to a given value, using an acid or a base; use a graduated Mohr pipet to deliver amounts of liquid accurate to 0.01 ml; read a Centigrade thermometer accurate to the nearest degree; and sterilize equipment using dry heat or an autoclave. Upon completion of this module, students will be able to perform the following function: use a job aid to identify all the equipment and reagents needed to collect stream samples for water quality analyses; clean and prepare sampling bottles needed for collecting stream samples for water quality analyses; and select the most suitable location for collecting water samples at a predesignated sampling site at a stream and take a grab sample.

Morris, W.F. and Ames, H.S., "Automation of the National Water Quality Laboratories, U.S. Geological Survey. I. Description of Laboratory Functions and Definition of the Automation Project", 1977, Lawrence Livermore Laboratory, Livermore, California.

In January 1976, the Water Resources Division of the U.S. Geological Survey asked Lawrence Livermore Laboratory to conduct a feasibility study for automation of the National Water Quality (NWQ) Laboratory in Denver, Colorado (formerly Denver Central Laboratory). Results of the study were published in the Feasibility Study for Automation of the Central Laboratories, Lawrence Livermore Laboratory, Rept. UCRL-52001 (1976). Because the present system for processing water samples was found inadequate to meet the demands of a steadily increasing workload, new automation was recommended. In this document details necessary for future implementation of the new system are presented, as well as descriptions of current laboratory automatic data processing and analytical facilities to better define the scope of the project and illustrate what the new system will accomplish. All pertinent inputs, outputs, and other operations that define the project are shown in functional designs.

National Field Investigations Center, "Comparison of Manual (GRAB) and Vacuum Type Automatic Sampling Techniques on an Individual and Composite Sample Basis", EPA/330/1-74/001, 1974, Denver, Colorado.

The National Field Investigations Center-Denver (NFIC-D) has been engaged in water quality and waste source evaluation studies since its inception. Due to the magnitude of the surveys, NFIC-D often has relied upon automatic samplers, particularly the SERCO, to collect samples of the wastewater streams. These samplers are used to collect water samples over specified periods for subsequent compositing; individual grab samples are collected manually. With the advent of compliance monitoring, the use of automatic samplers is expected to increase. Data provided by the manufacturer show that the automatic and manual sampling methods are equivalent. To confirm that the sampling methods are equivalent, NFIC-D conducted studies at a local wastewater treatment plant and statistically evaluated the results.   This paper presents the results of these studies.

Otson, R. et al., "Effects of Sampling, Shipping, and Storage on Total Organic Carbon Levels in Water Samples", Bulletin of Environmental Contamination and Toxicology, Vol. 23, No. 3, Oct. 1979, pp. 311-318.

The effect of sampling, shipping, and storage on TOC levels of water samples ranging from 400 mug/l to 7,500 mug/l in organic C content was investigated.   Standard solutions with calculated concentrations of 426, 2,042, 3,192, and 8,062 mug/l TOC were analyzed. The precision of TOC measurements was determined, and the experimental mean values and the calculated concentrations were used for regression analysis.   The precision of replicate measurements for the linearity study was less than 2 percent relative standard deviation (RSD) except in the case of the 426 mug/l TOC standard solution.   Linear regression lines with correlation coefficients greater than 0.999 were obtained for the raw water and the standard solutions.   An increase in TOC (46 mug/l) and TSD (7 percent) was observed for low TOC samples after bottling on day 0, possibly due to contamination during the bottling process.   Neither storage time nor storage temperature had any significant effect on the TOC values of tap water samples.   Storage of samples at 4°C causes a slight increase (approximately 50 mug/l) in TOC levels which can be detected in low TOC water samples but which is not noticeable in tap and river water samples.   Simulated shipping and subsequent storage of tap and river water samples showed no significant effect on TOC values.

Reckhow, K.H., "Techniques for Exploring and Presenting Data Applied to Lake Phosphorus Concentration", Canada Fisheries Research Board Journal, Vol. 37, No. 2, Feb. 1980, pp. 290-294.

Water quality sampling and data analysis are undertaken to acquire and convey information.   Therefore, when data are presented, the form of this presentation should be such that information transfer is high.   A technique is presented (that was developed for exploratory data analysis purposes) that can be used to display several sets of data on a single graph, indicating median, spread, skew, size of data set, and statistical significance of the median.   This technique is useful in the study of P concentration variability in lakes.   Additions to, and modifications of, this procedure are easily made and will often enhance the analysis of a particular problem.   Some suggestions are made for useful modifications of the plots in the study and for the display of P lake data and models.

Sanders, T.G. and Adrian, D.D., "Sampling Frequency for River Quality Monitoring", Water Resources Research, Vol. 14, No. 4, Aug. 1978, p. 569-576.

Sampling frequency for a water quality monitoring network is presented and applied to the Massachusetts portion of the Connecticut River basin.   The proposed frequency criterion is based on the assumption that the primary objectives of future river quality monitoring networks are the determination of ambient water quality conditions and an assessment of yearly trends rather than detection of stream or effluent standards violations.   The sampling frequency criterion is derived as a function of the random variability of the river flow.   The criterion is specifically related to the magnitude of the expected half width of the confidence interval of the mean of the random component on the annual statistic--mean log river flow. The appropriate sampling intervals are determined by specifying equality of this confidence interval half width, which insures a uniform reliability of the annual statistic.

Schock, M.R. and Schock, S.C., "Effect of Container Type on pH and Alkalinity Stability", Water Resources, Vol. 16, No. 10, 1982, pp. 1455-1464.

Errors in pH measurement of 0.1-0.2 units can propagate large errors in the predicted aqueous speciation. When accurate on-site analyses are impossible to obtain, changes in pH and alkalinity between the time of sampling and after transport to a laboratory must be minimized in many sampling programs. Several container and cap combinations were evaluated for their ability to preserve the pH and alkalinity of test waters that were out of equilibrium with atmospheric carbon dioxide over time periods representative of mail or other surface transport. Provided that no air bubbles were present when the sample bottle was filled and sealed, polyvinyl chloride bottles with caps having a conical polyethylene insert yielded the most satisfactory results. High density linear polyethylene bottles were also generally satisfactory. Cap design was observed to be very important. Guidelines for analysis and preservation are given, and if followed, laboratory measurements can be reliably substituted for field measurements in many instances.

Shnider, R.W. and Shapiro, E.S., "Procedures for Evaluating Operations of Water Monitoring Networks", EPA/600/4-76/050, 1976, U.S. Environmental Protection Agency, Las Vegas, Nevada.

This report is designed as a manual to evaluate the efficiency of surface-water quality monitoring networks whose primary objective is to document compliance with or progress toward attaining ambient water quality standards. The manual provides methods to evaluate the efficiency of each of seven operational areas: Network Plan and Design; Personnel; Facilities and Equipment; Sampling Quality Assurance; Data Distribution and Dissemination; and Agency Interactions. A technique is presented for the overall integrated evaluation of the operational areas. A final section provides methods to evaluate the efficiency of budgetary allocations.

Tuschall, J.R., Jr., Viborsky, R. and Surles, T., "Continuous Flow Monitoring Using Fluorescent Tracers", Pollution Engineering, Vol. 9, No. 10, Oct. 1977, p. 59.

The dilution method of monitoring flow by the tracer technique requires neither head loss nor interruption of flow. The method uses a continuous injection of tracer at a constant known rate, and the discharge through a pipe can be determined based on a mass balance equation for tracer concentration. Rhodamine WT and other fluorescent dyes allow accurate measurement over several orders of magnitude. The concentration of the fluorescent chemical is determined by measuring the intensity of the emitted radiant energy with a fluorometer. After a dye's stability in a sample of the effluent is checked, a point in the pipe network at least 100 pipe diameters upstream from the outfall is selected for dye injection. A calibrated pump wth an injection accuracy of at least

1 percent is used to feed the dye solution. Monitoring is done by a battery-operated field fluorometer with a flow-through cell, or a portable automatic water sampling system may be used to obtain effluent samples for fluorometer analysis in the laboratory.

Williams, T.J., "An Evaluation of the Need for Preserving Potable Water Samples for Nitrate Testing", Journal of the American Water Works Association, Vol. 71, No. 3, Mar. 1979, pp. 157-160.

All NO3- preservation methods attempt to slow or stop the biological processes of nitrification or denitrification which could result in changes in NO3- content. To test NO3- preservation, stored water samples were analyzed by the automated Cd reduction method for combined NO3--NO2-. No trend toward differences in test values was found among different storage conditions over a 4 day period. To test the effect of added NH3 on NO3- content, 6 ground water samples were split, with 1/2 of each adjusted with ammonium chloride so that the initial concentration of NH4-N was 20 mg/l. Samples were tested for 13 days. Changes only slightly exceeded the variation for the standard and for the corresponding sample with no NH3 added. The need for acid preservation of potable water samples for NO3- testing was not adequately established, suggesting that such a requirement will not significantly improve water supplier's ability to protect public health. A precautionary limit on sample holding times of 14 days for unpreserved potable water samples was suggested. As it relates to water supply monitoring, a blanket water preservation requirement for NO3- testing is apparently a costly precaution unlikely to affect water supply practices.

APPENDIX C

QUALITY ASSURANCE PROGRAM PLANNING
(ABSTRACTED FROM U.S. ENVIRONMENTAL PROTECTION AGENCY, 1980)

Quality assurance (QA) refers to the total integrated program for assuring reliability of monitoring and measurement data. Quality control relates to the routine application of procedures for obtaining prescribed standards of performance in the monitoring and measurement process. Sections 1 through 11 below describe the essential elements required in any QA program plan and indicates a format to be used. All eleven elements should be considered and addressed.

SECTION 1.0: IDENTIFICATION OF LABORATORY SUBMITTING QUALITY ASSURANCE PROGRAM PLANS

Each QA program plan should have a cover sheet listing the following information:

o   Document Title

o   Unit's Full Name and Address

o   Name, Address (if different from above), and Telephone Number of the Individual Responsible for the Unit

o   QA Officer

SECTION 2.0: INTRODUCTION

In this section, a brief background, purpose, and scope of the QA program plan is set forth. This section must state the rationale for QA application to monitoring and measurement activities and establish a basis for integrating the mandatory QA initiative into the program plan.

SECTION 3.0: QUALITY ASSURANCE POLICY STATEMENT

This policy statement provides the framework within which a unit develops and implements its QA program. The policy statement must describe the unit's goals and specify those requirements and activities needed to realize these goals. Policy is to be directed towards a formal commitment of time and resources necessary to ensure that data are as precise and accurate and as complete and representative as required. The policy statement indicates management's commitment to QA throughout the data generating and processing operations. All reported data must be accompanied by a calculation of precision and accuracy. Where approprite, a statement on the completeness, representativeness, and comparability is to be included.

SECTION 4.0: QUALITY ASSURANCE MANAGEMENT

4.1   Introduction

This section should present the QA management structure of the organization. This section of the plan should show the interrelationships between the functional units and the subunits which generate or manage data.

4.2   Assignment of Responsibilities

The office with overall QA responsibility for the organization must be identified. The success of the QA program depends on the QA office being organizationally independent of data-generating groups. Quality control (QC) responsibility within the various organizational subunits should be stated.

4.3   Communications

An "organizational type" chart should be used to indicate the internal and external flow of QA information and reports. The flow of reports and other QA information should be addressed and should specify the following: (1) the responsibility of the QA office; (2) what levels of the unit will receive various reports; (3) how often the reports will be issued; (4) who will review these reports; and (5) who will take follow-up action, if necessary.

4.4   Document Control

QA reports and other vital information, plans, and directives should be maintained under a document control procedure. Distribution lists of personnel who need to receive QA reports and information should be maintained as part of the document control system.

4.5   QA Program Assessment

This section should describe how an organization will continually review its QA program's benefits and effectiveness.

SECTION 5.0: PERSONNEL QUALIFICATIONS

Each organization must assure that all personnel performing tasks and functions related to data quality have the needed education, training, and experience.   This includes laboratory technicians, analysts, maintenance technicians, supervisors, principal investigators, statisticians, project officers, and QA staff. Personnel qualifications and training needs should be identified.

SECTION 6.0: FACILITIES, EQUIPMENT, AND SERVICES

In this section, the unit's approach to the selection, evaluation, and maintenance (both routine and corrective) of its equipment, facilities, and services should be described.   The QA program plan should address general environmental aspects of the facilities and equipment which might impact data quality (temperature, humidity, lighting, dust levels, etc.).   It should also consider maintenance requirements for general utilities and housekeeping services (voltage control, reagent purity, supply of air and water, refrigerators, incubators, laboratory hoods, glove boxes, etc.).   In addition, this plan should identify the monitoring and inspection procedures, service manuals, maintenance procedures, and service agreements utilized to maintain performance standards for equipment, including essential ancillary equipment.

SECTION 7.0: DATA GENERATION

This section of the QA program plan should describe procedures to assure the generation of reliable data by all environmentally related measurement activities. To assure that all data generated are scientifically valid, defensible, comparable, and of known precision and accuracy, these plans must provide for the use of QA project plans and standard operating procedures (SOPs).

7.1   Quality Assurance Project Plans

Each QA program plan must identify the types of monitoring and measurement activities for which detailed QA project plans must be developed and followed. QA project plans should provide for the review of all activities which could directly or indirectly influence data quality and the determination of those operations which must be covered by SOPs. Activities which should be reviewed include:

o   General Network Design

o   Specific Sampling Site Selection

o   Sampling and Analytical Methodology

o   Probes, Collection Devices, Storage Containers, and Sample Additives or Preservatives

o   Special Precautions, such as heat, light, reactivity, combustability, and holding times

o   Reference Procedures

o   Instrument Selection and Use

o   Calibration and Standardization

o   Preventive and Remedial Maintenance

o   Replicate Sampling

o   Blind and Spiked Samples

o   Colocated Samplers

o   QC Procedures, such as intralaboratory and intrafield activities and interlaboratory and interfield activities

o   Documentation

o   Sample Custody

o   Transportation

o   Safety

o   Data Handling Procedures

o   Service Contracts

o   Measurement of Precision, Accuracy, Completeness, Representa-
     tiveness, and Comparability

7.2   Standard Operating Procedures

SOPs should be developed and used to implement routine QC requirements
for all monitoring programs, repetitive tests and measurements, and for
inspection and maintenance of facilities, equipment, and services.

SECTION 8.0: DATA PROCESSING

QA program plans should summarize how all aspects of data processing
will be managed and separately evaluated in order to characterize and maintain
the integrity and quality of the data.   Data processing includes collection,
validation, storage, transfer, and reduction.

8.1   Collection

The QA project plan must address both manually collected data and
computerized data acquisition systems. The internal checks that must be used
to avoid errors in the data collection process should be identified.

8.2   Validation

Data validation is defined as the process whereby data are accepted or
rejected based on a set of criteria.   This aspect of QA involves establishing
specified criteria for data validation.

8.3   Storage

At every stage of data processing at which a permanent collection of data
is stored, procedures must be established to ensure data integrity and security.
QA project plans should indicate how specific types of data will be stored with
respect to media, conditions, location, retention time, and access.

8.4   Transfer

QA project plans should describe procedures which must be used to ensure
that data transfer is error-free (or have an admissable error rate), that no
information is lost in the transfer, and that the input is completely recoverable
from the output. Examples of data transfer are: copying raw data from a
notebook onto a data form for keypunching, converting a written data set to
punched cards, or copying from computer tape to disk. A QA project plan
should indicate a minimum number of data transfer steps in the data processing.

8.5   Reduction

Data reduction includes all processes which change either the form of expression or quantity of data values or numbers of data items. This category includes validation, verification, and statistical and mathematical analysis. It is distinct from data transfer in that it entails a reduction in the size or dimensionality of the data set.

SECTION 9.0: DATA QUALITY ASSESSMENT

QA program plans should describe how all data generated will be assessed for accuracy, precision, completeness, representativeness, and comparability. The QA program should require that data be accompanied by a calculation of precision and accuracy. Where appropriate, a statement on the completeness, representativeness, and comparability also should be included.

SECTION 10.0: CORRECTIVE ACTION

Each QA program plan should include provisions for establishing and maintaining QA reporting or feedback channels to the appropriate management authority to ensure that early and effective corrective action can be taken when data quality falls below required limits. The QA program plan should also include provisions to keep responsible management informed of the performance of all data collection systems and should describe the mechanism(s) to be used when corrective actions are necessary. Corrective action relates to the overall QA management scheme: who is responsible for taking corrective actions; when are corrective actions to be taken; and who follows up to see that corrective actions have been taken and that they have produced the desired results.

SECTION 11.0: IMPLEMENTATION REQUIREMENTS AND SCHEDULE

The QA program plan should be accompanied by an implementation schedule.

SELECTED REFERENCE

U.S. Environmental Protection Agency, "Guidelines and Specifications for Preparing Quality Assurance Program Plans", QAMS-004/80, Sept. 1980, Washington, D.C.

APPENDIX D

CASE STUDIES OF WATER QUALITY MONITORING

Burge, R.E., Herrmann, R. and Matthews, R.C., Jr., "Remote Sensing of Water Quality and Weather in Great Smoky Mountains National Park: Phase I", R/RM-30, 1979, U.S. Forest Service, Atlanta, Georgia.

Four self-contained Convertible Data Collection Platforms (CDCP's) have been developed and deployed as an integral part of a comprehensive water quality and environmental monitoring program for remote natural areas at Great Smoky Mountains National Park. Multiple parameter sensor packages regularly monitor (1) water quality parameters (dissolved oxygen, hydrogen ion concentration (pH), oxygen reduction potential, conductivity, and temperature), and (2) meteorological parameters (wind direction and speed, ambient temperature, relative humidity, barometric pressure and cumulative precipitation). The monitoring system is comprised of the multi-parameter sensor packages, the CDCP unit, a transmitting antenna, and a power source.

Raytheon Company, "Environmental Monitoring Report, United States Department of Energy, Oak Ridge Facilities", Y/UB-16, 1982, Washington, D.C.

The Environmental Monitoring Program for the Oak Ridge area includes sampling and analysis of air, water from surface streams, creek sediments, biota, and soil for both radioactive and nonradioactive materials. This report presents a summary of the results of the program for calendar year 1981. Surveillance of radioactivity in the Oak Ridge environs indicates that atmospheric concentrations of radioactivity were not significantly different from other areas in East Tennessee. Concentrations of radioactivity in the Clinch River and in fish collected from the river were less than 2 percent of the permissible concentration and intake guides for individuals in the offsite environment. While some radioactivity was released to the environment from plant operations, the concentrations in all of the media sampled were well below established standards. Surveillance of nonradioactive materials in the Oak Ridge environs shows that established limits were not exceeded for those materials possibly present in the air as a result of plant operations. The chemical water quality data in surface streams obtained from the water sampling program indicated that average concentrations resulting from plant effluents were in compliance with state stream guidelines with the exception of fluoride at monitoring Station B-1 which was 120 percent of the guideline. National Pollutant Discharge Elimination System (NPDES) permit compliance information has been included in this report. During 1981 there were no spills of oil and/or hazardous materials from the Oak Ridge installations reported to the National Response Center.

Shackelford, W.M. and Keith, L.H., "Frequency of Organic Compounds Identified in Water", EPA/600/4-76/062, 1976, U.S. Environmental Protection Agency, Athens, Georgia.

This study was initiated for the purpose of compiling a list of all organic compounds that have been found in water. This report contains the names of compounds found, their location or a reference to a published study, the type of water in which they are found, and the date of sampling or report generation. About 5700 entries are in this list at present. Among them are 1259 different compounds that have been found in one or more of 33 different water types. Tables in the text include lists of the data file sorted by compound name, location or reference, and water type. Also included are tables that summarize the frequency of occurrence of each compound, location or reference, and water type. No attempt has been made to include data from studies that include only analysis for specific compounds. The list is meant to include, however, all survey-type analyses of water samples.

U.S. Environmental Protection Agency, "Model State Water Monitoring Program", EPA/440/9-74/002, 1975, Washington, D.C.

This model state water monitoring program was developed by a panel of Federal and state professionals actively engaged in managing and operating monitoring programs. It is presented to others in monitoring and the field of water pollution control in order to: (1) provide some basis to the States for building and operating water monitoring programs; (2) illustrate the various types of monitoring activities, their costs and their uses; and (3) suggest to EPA Regions and States how they can best use monitoring resources in carrying out their responsibilities in pollution control and abatement.

U.S. Environmental Protection Agency, "Assessment of the Region 5 Water Monitoring Activities", EPA-905/4-81-001, 1981, Chicago, Illinois.

A review has been completed of the water monitoring activities within the Region to determine the adequacy of existing monitoring activities to meet program needs. Following a determination of program needs and an identification of the programs and program objectives established to satisfy those needs, an assessment was made of both the State and Federal activities against the program objectives. Evolving issues, primarily the need to develop monitoring capability for toxics and hazardous materials were also reviewed. In addition, the changing direction in resource utilization evident from strategy papers under development by Headquarters have also been taken into consideration. From this assessment and review, findings and recommendations were formulated which lead to a monitoring management strategy which will result in the coordination of resource utilization and the re-direction of monitoring resources needed to ensure a meaningful data base for use by program managers during the period 1981-1985. The findings, recommendations and strategy are described in Chapters 1 and 2. The supporting information on program needs, monitoring program objectives and the assessment of current programs are described in the Appendices. The base year for the review was CY 1979.

APPENDIX E

ANALYTICAL TECHNIQUES FOR WATER QUALITY PARAMETERS

Baltisberger, R.J., "The Microdetermination of Mercury Species in Natural Water Systems by Liquid Chromatography", WI-222-010-75, 1975, Washington, D.C.

Mercury in environmental water samples can exist in several cationic forms such as $Hg(2+)$, $Hg2(2+)$ and $CH3Hg(+)$. Methylmercury cation, $CH3Hg(+)$ is formed by the methylation of $Hg(2+)$ by aquatic microorganisms. Metallic mercury has a solubility of 0.0000003 moles/liter (60 ppb) at 25C and is in equilibrium with $Hg(2+)$. This study has shown that this equilibrium is slow to be established. The fate and form of mercury compounds in aqueous solution involves complex interactions. In order to adequately establish water quality standards for this element, it is desirable to have analytical techniques capable of differentiation of the exact cationic forms of mercury in environmental water samples. Two analytical techniques were developed and tested for (1) the measurement and differentiation of inorganic and organic mercury cations in environmental samples and (2) the separation by ion exchange chromatographic means of these cations and their ultimate analysis by flameless atomic spectrophotometry. The first method is useful in the concentration range from 0.1 to 10 ppb $Hg2(2+)$, $Hg(2+)$ or $CH3Hg(+)$. The latter method is useful from 1 to 10 ppb mercury.

Baltisberger, R.J., "The Differentiation of Inorganic and Organomercury Species in Aqueous Samples", WI-221-041-76, 1977, Washington, D.C.

Mercury in environmental water samples can exist in several forms such as $Hg(+2)$, Hg, and $CH3Hg(+)$. Methylmercury cation, $CH3Hg(+)$ is formed by the methylation of $Hg(+2)$ by aquatic microorganisms. Metallic mercury has a solubility of 3×10 to the -7th power moles/liter (60 ppb) at 25 degrees and is in equilibrium with $Hg(+2)$ and $Hg2(+2)$. The fate and form of mercury compounds in aqueous solution involve complex interactions. In order to adequately establish water quality standards for this element, it is desirable to have analytical techniques available capable of differentiation of the exact forms of mercury in environmental water samples. An analytical technique was developed and tested for the measurement and differentiation of inorganic and organic mercury cations in environmental samples and their ultimate analysis by flameless atomic spectrophotometry. The method is useful in the concentration range from 10 to 100 ppb $Hg2(+2)$, $Hg(+2)$, or $CH3Hg(+)$.

Bishop, C.T. et al., "Anion Exchange Method for the Determination of Plutonium in Water: Single-Laboratory Evaluation and Interlaboratory Collaborative Study", EPA/600/7-78/122, 1978, U.S. Environmental Protection Agency, Las Vegas, Nevada.

This report gives the results of a single-laboratory evaluation and an interlaboratory collaborative study of a method for determining plutonium in water. The method was written for the analysis of 1-liter samples and involved   coprecipitation,   acid   dissolution,   anion   exchange, electrodeposition, and alpha pulse-height analysis. The complete method is given in the first appendix to the report. After the single-laboratory evaluation of the selected method, four samples were prepared for the collaborative study. There were two river water samples, a substitute ocean water sample, and a sample containing sediment. These samples contained plutonium-239 and plutonium-238 at concentrations ranging from 0.42 to 28.9 dis/min/liter. Standard deviations of the collaborative study plutonium concentrations ranged from 5 percent to 13 percent. In three cases standard deviations agreed with what was expected from counting statistics. It is believed that hydrolysis occurred in the river water samples resulting in errors greater than what was expected from counting statistics.

Chakravorty, R. and Van Grieken, R., "Co-precipitation with Iron Hydroxide and X-Ray Fluorescence Analysis of Trace Metals in Water", International Journal of Environmental Analytical Chemistry, Vol. 11, No. 1, 1982, pp. 67-80.

Preconcentration of transition trace ions by coprecipitation on iron-hydroxide has been combined with energy-dispersive X-ray fluorescence for environmental water analysis. The optimized preconcentration procedure implies adding 2 mg of iron to a 200 ml water sample, adding dilute NaOH up to pH 9, filtering off on a nuclepore membrane after a 1 hour equilibration time, and analyzing. Quantitative recoveries could then be obtained for Ni, Cu, Zn and Pb, e.g. at the 10 mu g/l level in waters of varying salinity while Mn partially collected. The precision is 7-8 percent at the 10 mu g/l level, and the detection limits are in the 0.5-1 mu g/l range. Various environmental water samples are analyzed by way of illustration.

Chatfield, E.J., Dillon, M.J. and Stott, W.R., "Development of Improved Analytical Techniques for Determination of Asbestos in Water Samples", PB83-261651, 1983, National Technical Information Service, U.S. Department of Commerce, Springfield, Virginia.

Sample preparation techniques were examined for the analysis of asbestos fiber concentrations in water. The carbon-coated Nuclepore technique using a polycarbonate filter proved to be superior to either the "drop" or the collapsed membrane filter technique. Compared with plasma ashing, ozone-ultraviolet light oxidation of water samples was found to be a simpler and superior technique for removal of organic materials. Experiments revealed that large proportions of the suspended asbestos fibers could become attached to the inside surfaces of sample containers. This effect was caused by trace organic materials of bacterial origin. Ozone oxidation, carried out inside the collection container, released the attached fibers into the water again. If bacteria and their products were excluded initially, and if absolute sterility was maintained thereafter, suspensions of both chrysotile and crocidolite appeared to be stable for long periods of time. Tests of reference suspensions in sealed glass ampoules stored for almost two years produced fiber concentration values statistically compatible with those obtained at the time of sample preparation.

Cook, P.P., Duvall, P.M. and Bourke, R.C., "Improved Methods of Oil-in-Water Analysis", Water and Sewage Works, Apr. 1978.

The standard reference method used to determine oil and grease levels in water samples is based on the gravimetric analysis of a solvent extract. The precision and accuracy of this method is significantly improved by the following modifications: increasing the quantity of extracting Freon solvent; extracting for 45 min with mechanical stirring; and eliminating moisture adsorption on the gravimetric flask. A semiautomatic Horiba OCMA-200 oil analyzer performs as well as the modified reference method, if care is taken to use the proper extraction time, prevent instrument overheating, and acidify the sample just prior to its injection.

Crowther, J., "Autoclave Digestion Procedure for the Determination of Total Iron Content of Waters", Analytical Chemistry, Vol. 50, No. 4, Apr. 1978, pp. 658-659.

Thirty ml of water sample and 2.0 ml of digestant acid (640 ml of hydrochloric acid plus 30 gr of hydoxylammonium chloride/l) were autoclaved at 121ºC for 1 hr. After cooling to room temperature, a colorimetric test, using the chromophore 2,4,6-tri(2-pyridyl)-1,3,5-triazine (TPTZ) was made. To evaluate the autoclave-TPTZ procedure, the Fe content of 610 samples was determined by the subject procedure and by an acid-digestion method recommended by Standard Methods. Samples were divided into 3 groups--those from domestic water supplies and landfill leachates, samples collected from rivers and lakes during winter, and river and lake samples collected during the spring runoff. Linear regression analyses for the 3 sets of data indicate that the 2 procedures give comparable results. The lowest SD was obtained from the first group which contained little or no particulate matter, and the largest SD was associated with the spring runoff samples which contained so much silt that it was difficult to select a representative sample. Winter samples contained small amounts of particulates and the SD was intermediate. If the autoclave digestion technique was inadequate, the measured Fe content would tend to be lower than that determined by the Standard Methods procedure. Such a bias (5-10 percent lower) was found when creek and snow samples from 2 locations were analyzed. Although this weakness can be tolerated in view of the infrequent occurrence and the size of the error involved, analysts should evaluate the technique for their particular waters before adopting it.

Crowther, J., "Semiautomated Procedure for the Determination of Low Levels of Total Manganese", Analytical Chemistry, Vol. 50, No. 8, July 1978, pp. 1041-1043.

Total Mn is determined in water samples using batch digestion by autoclaving in acid medium followed by automated colorimetry with a Technicon AutoAnalyzer sampling system. The colorimetric procedure for Mn is based on its formation in a formaldoxime complex, which produces a pink color that has a constant intensity in the pH range 9.0-10.5 and is absorbed in the 480-nm range. Cation interference is suppressed by complexing agents while color interference is eliminated by using a blanking stream which is synchronized with the color stream. Both streams receive identical quantities of reagents, but the order of addition is altered to prevent the formation of the Mn-formaldoxime complex in the blank stream. In the range of application (0.006-0.200 mg/l Mn), the accuracy and precision of the method are comparable to those of flame AAS.

Downes, M.T., "An Improved Hydrazine Reduction Method for the Automated Determination of Low Nitrate Levels in Freshwater", Water Research, Vol. 12, No. 9, 1978, pp. 673-675.

An automated nitrate (NO3-) determination is described in which NO3- is reduced to nitrite with hydrazine sulfate under alkaline conditions in the presence of Cu+2 and Zn+2. Interferences encountered in natural water samples were eliminated by the addition of Zn+2 to the Cu+2 catalyst solution. The method is suitable for the determination of low NO3-N concentrations and compares favorably with the manual copperized Cd technique for freshwater samples containing 10-800 mg/m3 NO3-N. The method is also linear at NO3- concentrations below 10 mg N/m3. The SDs of blanks and of samples containing 2 mg NO3-N/m3 were 0.013 and 0.06 mg N/m3, respectively, at an analysis rate of 30 samples/hr.

Elsenreich, S.J. and Hullett, D.A., "Determination of Free and Bound Fatty Acids in River Water by High Performance Liquid Chromatography", Analytical Chemistry, Vol. 51, No. 12, Oct. 1979, pp. 1953-1960.

A technique is described for the isolation and quantitation of free and bound fatty acids (FA) in river water. The method involves sequential liquid-liquid extraction of the water sample by 0.1N HCl, benzene:methanol (7:3) and hexane:ether (1:1). The resultant extract was concentrated, and the fatty acids were separated as a class on Florisil using an ether:methanol (1:1) and (1:3) elution. Final determination of individual fatty acids was accomplished by forming the phenacyl ester and separating by HPLC. Results given for the distribution of FA in Mississippi River water show the general applicability of the technique to complex environmental samples.

Favretto, L., Stancher, B. and Tunis, F., "Improved Method for the Spectrophotometric Determination of Polyoxyethylene Nonionic Surfactants in the Presence of Cationic Surfactants", International Journal of Environmental Analytical Chemistry, Vol. 14, No. 3, 1983, pp. 201-214.

A new method, free from interference of cationic surfactants, is proposed for the determination of polyoxyethylene nonionics in waters at ultra-trace levels. The procedure involves (1) a pre-extraction from water sample with dichloromethane, and (2) the determination of nonionics as potassium picrate active substances (PPAS). In the aqueous phase the polyether chain reacts reversibly with potassium cations (in large excess) to form a positive complex, which is extracted and concentrated in 1,2-dichloroethane (1,2-DCE) through ion pairing with the chromophoric picrate. The yellow color of ion pairs extracted into the organic layer disappears immediately by extraction with water, which lowers drastically the concentration of reagents. The difference between the absorbances gives the PPAS attributable to polyoxyethylene nonionics only, because PPAS due to cationics are not extractable. The method was tested in sea waters polluted by urban liquid wastes.

Johnston, J.B. and Herron, J.N., "A Routine Water Monitoring Test for Mutagenic Compounds", UICU-WRL-79-0141, 1979, University of Illinois, Champaign, Illinois.

The development of a simple, relatively comprehensive method for the recovery of nonvolatile mutagenic compounds from surface waters is described. The method recovers compounds by sequential passage of the water sample through a silica gel bed, then a cation-exchange bed, and then an anion-exchange bed--all contained in a single multi-bed column of glass and teflon, the parfait column. Nonvolatile compounds not adsorbed to any of these beds were recovered following concentration of the column effluent by vacuum distillation at less than 30 degrees. The beds of the parfait column were separated and eluted independently. Using the Ames Salmonella/microsome reversion assay, each residue was assayed for mutagenic activity. The method was evaluated by recovery of five known mutagens, benzo(a)pyrene, 4-nitroquinoline-1-oxide, ethidium bromide, nitrofurylfuramide, and sodium azide, each initially spiked into a sample of laboratory deionized water and an environmental water sample to a final concentration of less than 3 ppb. The method has been used to survey ten Illinois surface waters for naturally occurring mutagenic activity. The parfait/distillation method differs from other techniques for the recovery of waterborne mutagens in its emphasis on the recovery of nonvolatile compounds and neutral water soluble compounds. This method has also detected a significant mutagenic activity in samples as small as 2 gallons of water, a volume consumed by a normal person every few days.

Kloosterboer, J.G. and Goossen, J.T.H., "Determination of Phosphates in Natural and Waste Waters After Photochemical Decomposition and Acid Hydrolysis of Organic Phosphorus Compounds", Analytical Chemistry, Vol. 50, No. 6, May 1978, pp. 707-711.

Total dissolved and suspended phosphate in water samples may be determined after photochemical decomposition of organic phosphorus compounds and thermal hydrolysis of acid-hydrolyzable phosphates, followed by conventional spectrophotometric determination of the liberated orthophosphate as molybdenum blue. A 75-W medium pressure Zn-Cd-Hg lamp is used for 20-25 min photolysis and hydrolysis. The combined action of UV radiation and heat from the lamp simultaneously converts organic phosphates and acid-hydrolyzable phosphates to orthophosphate for the measurement of total phosphate. A thin Al sheet placed between the lamp and acidified sample solution allows only hydrolysis to occur, so that specific levels of orthophosphate and ortho + acid-hydrolyzable phosphate can be determined. The method avoids complicated and time-consuming chemical pretreatment and may easily be automated.

Lantz, J.B., Davenport, R.J. and Wynveen, R.A., "Evaluation of Breadboard Electrochemical TOC/COD Analyzer: Advanced Technology Development", LSI-TR-310-4-3, 1980, Cleveland, Ohio.

The Advanced Breadboard TOC/COD Analyzer has been designed, fabricated, assembled and the testing program for its evaluation has been initiated. The Analyzer has been designed for simple, automated operation. It incorporates many advanced features that are designed to either provide more data than is available using other analyzers, or to more effectively integrate the Analyzer into process monitoring/control applications and applications involving water quality monitoring in remote areas. Further activities will include completion of the testing program. Also, a method of preventing interferences in the organic solute measurements due to chloride in the water samples will be developed and integrated into the Advanced Breadboard.

Manahan, S.E. et al., "An Analytical Method for Total Heavy Metal Complexing Agents in Water and Its Application to Water Quality Studies", W74-02658, 1973, University of Missouri, Columbia, Missouri.

This report describes the development of several methods of analysis for low levels of complexing agents, particularly chelating agents. The first method developed was an atomic absorption analysis of strong heavy metal chelating agents. This method is based upon the fact that when copper ion is added to a water sample and the pH adjusted to 10, the only copper that remains in solution is that which is in a complexed or chelated form. The method was used on a number of natural water samples. The method was extended to the analysis of cyanide ion. A new method was developed in which the copper is solubilized from a copper-containing chelating ion exchange resin. This method is much more rapid than the first method described, and it is applicable to automated procedures and as a detection system for chelating agents separated by liquid chromatography.

Manning, D.C., "Water/Wastewater Analysis by Atomic Absorption Spectroscopy", Water and Sewage Works, 1979.

Atomic absorption spectroscopy (AAS) can measure major water components such as Ca, Mg, and Na and trace elements such as Hg at concentrations less than 0.0001 mg/l. In the nebulizer-burner flame system, the water sample is converted to a mist, mixed with fuel and oxidant gases, and burned, reducing the small dissolved particles to an atomic vapor. When the sample has been atomized, the sample atoms in the flame can absorb light emitted from a hollow cathode lamp at specific wavelengths, and the concentration of the specific atom is determined by the amount of light absorbed. Instruments are available in single-beam and double-beam systems and should have a resolution of 0.2 nm. AAS can also be used to determine emission measurements but requires an operator of greater skill and experience than is normally necessary.

Marks, A., "Ion Selective Electrodes: Progress and Potential", Intech, Vol. 29, No. 6, 1982, pp. 9-10.

Ion selective electrodes (ISEs) are emerging in new configurations for increasingly varied applications. ISEs develop potentials related to the activity of ions in solution. Familiar pH meters are based on hydrogen ISEs; similar instruments can detect other ions in solution. Although concentration determinations from ion activity measurements are indirect, the instruments are faster and less expensive than most classical analyzers. Water monitoring is a prime application. These systems take advantage of sensitivity as well as response speed.

Matsumoto, G., "Comparative Study on Organic Constituents in Polluted and Unpolluted Inland Aquatic Environments II. Features of Fatty Acids for Polluted and Unpolluted Waters", Water Resources, Vol. 15, No. 7, 1981, pp. 779-787.

Fatty acids were analyzed for polluted river waters from the Tokyo area and unpolluted river, brook, reservoir and pond waters from the Ogasawara (Bonin) Islands to elucidate their features for polluted and unpolluted waters. Fatty acids ranging from the carbon chain length of C sub(8)-C sub(34) including unsaturated and branched acids were found with the great predominance of even-carbon numbers and lower molecular weight ranges (C sub(13)-C sub(19)) in the water samples from the Tokyo area and Ogasawara Islands. It was thus confirmed that no marked changes in fatty acid composition between polluted and unpolluted waters are absent.

National Environmental Research Center, "Methods for Chemical Analysis of Water and Wastes", EPA-16020-07/71, 1971, U.S. Environmental Protection Agency, Cincinnati, Ohio.

This handbook describes chemical analytical procedures to be used in Water Quality Office (WQO) laboratories. Except where noted under "Scope and Application" for each constituent, the methods can be used for the measurement of the indicated constituent in both water and wastewaters and in both saline and fresh water samples. Instrumental methods have been selected in preference to manual procedures because of the improved speed, precision, and accuracy. Procedures for the Technicon AutoAnalyzer have been included for laboratories having this equipment available.

Rock, J.C. and Post, J.A., "Field Analysis of Ten Parts Per Billion Trichloroethylene in Water, Using a Portable, Self-Contained Gas Chromatograph", 14th Annual Conference on Trace Substances in Environmental Health, 1980, University of Missouri, Columbia, Missouri.

A procedure is described which permits rapid field analysis for trace amounts of trichloroethylene (TCE) in recreational or drinking quality water samples. The water sample is sealed into a container with an equal volume of headspace air and shaken for 60 seconds to establish equilibrium conditions. A small aliquot of headspace air is withdrawn and injected into a Century Systems Model 128 Organic Vapor Analyzer. This portable, self-contained, flame ionization gas chromatograph, equipped with an analytical column made from 10 percent SP 2100 on Supelcoport 60/80 mesh packing in 1/8-in aluminum or teflon tubing, provides adequate separation with a 24-in column length at 25°C or a 9-in column length at 0°C, and a 12 ml/min hydrogen carrier gas flow rate. The field analysis is based upon the height of the TCE peak, which elutes about 3 minutes after injection.

Salim, R. and Cooksey, B.G., "Effect of Centrifugation on the Suspended Particles of River Waters", Water Resources, Vol. 15, No. 7, 1981, pp. 835-839.

Samples of river water have been analyzed for their particle size distribution using the Coulter Counter. The feasibility of the application of the centrifugal method for particle size analysis on river water samples has been investigated. The efficiency of centrifugation for separating the colloidal and suspended particles from river water samples has been determined.

Seitz, W.R., "Evaluation of Flame Emission Determination of Phosphorus in Water", EPA-660/2-73-007, 1973, U.S. Environmental Protection Agency, Corvallis, Oregon.

A flame spectrometer for phosphorus analysis was evaluated. Response to phosphorus in the form of H3PO4 was linear from 3 micrograms/liter, the detection limit, to 120 mg/l, the highest concentration tested. Metal ions depress phosphorus emission and must be removed by cation exchange prior to analysis. High concentrations (equal or greater than mg/l) of sulfur interfere positively. Volatile phosphorus compounds produce a larger signal for a given phosphorus concentration than nonvolatile compounds. River water samples were spiked with inorganic and organic phosphorus and analyzed. The measured phosphorus concentrations were 10-25 percent lower in river water than in deionized water.

Sekerka, I. and Lechner, J.F., "Potentiometric Determination of Organohalides in Natural Water Using Tenax Adsorption and Combustion", International Journal of Environmental Analytical Chemistry, Vol. 11, No. 1, 1982, pp. 43-52.

The method described in this paper involves adsorption of organohalides from natural water samples on porous polymer Tenax GC followed by thermal desorption at 400°C and oxidative combustion to hydrogen halides. The halides are captured in water solution containing formaldehyde and sulfamic acid and measured either by direct potentiometry with high sensitive chloride ion selective electrode or by potentiometric titration with Hg super (2+) titrant solution. The efficiencies of the adsorption, combustion and detection as well as elimination of interferences are presented. Analysis time after adsorption is rapid (2 min) and procedures are simple, making the technique suitable for routine application.

Sharma, S.R., Rathore, H.S. and Ahmed, S.R., "New Specific Spot Test for the Detection of Malathion in Water", Water Resources, Vol. 17, No. 4, 1983, pp. 471-473.

This paper describes a new specific colorimetric spot test for the detection of malathion residues in water. Activated charcoal is used to recover and concentrate malathion from water samples. Ethanolic malathion solution is hydrolyzed with potassium hydroxide to give potassium fumarate which gives a red color on heating with acetic anhydride. The lower limit of detection is 1 mg/l.

Slabbert, J.L. and Morgan, W.S.G., "Bioassay Technique Using Tetrahymena Pyriformis for the Rapid Assessment of Toxicants in Water", Water Resources, Vol. 16, No. 5, 1982, pp. 517-523.

A bioassay technique utilizing the respiratory response of Tetrahymena pyriformis to intoxication was developed. Changes in oxygen uptake rate caused by a variety of toxicants produce measurable results within 10 min. Although not sufficiently sensitive to detect some chemicals at concentrations within the limits specific for drinking water, the technique should be more than adequate to monitor the quality of industrial effluents.

Stepanenko, V.E. and Muslova, N.M., "Chromatographic Determination of Organic Carbon in Aqueous Solutions", Zavodskaya Laboratoriya, Vol. 44, No. 9, Sept. 1978, pp. 1068-1071.

Concentrations of volatile organic carbon (VOC), inorganic carbon (IC), and nonvolatile organic carbon (NVC) in water were determined using an apparatus attached to a commercial GC with a flame ionization detector. The water sample was made alkaline to fix dissolved CO_2 and was introduced into a desorber where a carrier-gas flow extracted volatile organic substances which were oxidized in an oxidative reactor to CO_2. This CO_2 was hydrogenated to CH_4 detectable by the flame ionization detector and recorded as VOC. The sample in the absorber was then acidified with H_2SO_4 to remove the IC, which was recorded on the chart in the form of a second peak, representing the content of carbonates and CO_2 in the sample. The remaining NVC was fed by microsyringe into the oxidative reactor and recorded as a third peak, representing the NVC content. Minimum determinable concentrations of VOC, IC, and NVC were 0.002, 1.0, and 0.002 mg/l, respectively. Maximum impurities in waters investigated ranged from 200 to 250 mg/l. The method was used to analyze river water, outputs from biological installations, industrial effluents, and technological solutions of mineral products.

Thurnau, R.C., "Ion Selective Electrodes in Water Quality Analysis", EPA/600/2-78/106, 1978, U.S. Environmental Protection Agency, Cincinnati, Ohio.

The maintenance of water quality whether at the treatment plant or out in the distribution system is predicated on accurately knowing the condition of the water at any particular moment. Ion selective electrodes have shown tremendous potential in the area of continuous water quality analysis, and were employed by the Water Supply Research Division's Mobile Water Quality Laboratory to monitor: alkalinity, calcium, chloride, fluoride, hardness, nitrate, and pH. The pH and the chloride electrodes were housed in a commercial unit and linked to the computer with a minimum number of operating problems. The other parameters required more development and all relied on ionic strength or pH buffers to swamp out problems of activity and ionic strength. The test periods were usually about a week in length, and data were presented as to the reliability and accuracy of the electrodes. It was found that the electrodes performed quite well, and when compared to accuracy statistics found in Standard Methods for the Examination of Water and Wastewater, the electrode methods were in the same region.

Ullman, F.G., "Investigation of Laser Raman Spectroscopy for Analysis of Water Quality", W77-01817, 1976, U.S. Department of the Interior, Washington, D.C.

Laser Raman spectroscopy was studied as a tool for analysis of water quality. Lower limits for the detection of nitrates and sulfates in distilled water were determined to be 20 ppm and 8.5 ppm, respectively. The herbicides atrazine, picloram, trifluoralin, amiben and dicamba (primarily in methanol solutions as the solubility of most of these in water is very low) were also investigated. Atrazine could be detected down to 750 ppm (as atrazine). Trifluoralin absorbed most of the incident light at the two test laser frequencies so a spectrum could not be obtained and the spectra of the other herbicides were obscured by their own fluorescence. At the lower limit of detectability of nitrate and sulfate ions, the impurity signal is obscured by a "background" signal that can be as large as a few orders of magnitude above the intrinsic background of the instrument. These "background" signals are reported in the literature to be characteristic of liquid solvents. This unavoidable background forces a lower limit on the utilization of Raman spectroscopy for studying impurities in unconcentrated water samples. Further studies were made on concentrated samples, prepared by reversible ion exchange using an ion exchange resin.   Concentration by this method appears feasible. However, the reproducibility of removal and recovery of the desired ion has not been completely established.   The relative merits of this technique for studies of water quality are discussed.

Water Research Centre, "Simultaneous Multi-Element Analysis of Aqueous Solutions", WRL-RN-8, 1977, Stevenage, England.

Techniques are discussed for the simultaneous multi-element analysis of water samples, in order to determine trace metals. The four techniques surveyed meeting the requirements of multi-element analysis and using standard instrumentation are: spark-source mass spectroscopy, neutron activation analysis, X-ray fluorescence spectrometry, and optical emission spectrometry.

APPENDIX F

BACTERIOLOGICAL WATER QUALITY MONITORING

Berman, D., Rohr, M.E. and Safferman, R.S., "Concentration of Poliovirus in Water by Molecular Filtration", EPA-600/J-80-161, 1980, U.S. Environmental Protection Agency, Cincinnati, Ohio.

The efficiency of concentrating poliovirus 1 from distilled water samples was determined by using a recirculating-flow molecular filtration system. The most efficient recoveries were achieved against membranes with a 10,000 nominal molecular weight limit pretreated with flocculated beef extract. This procedure yielded a mean virus recovery of 67 percent.

Block, J.C. and Rolland, D., "Method for Salmonella Concentration from Water at pH 3.5, Using Microfiber Glass Filters", Applied and Environmental Microbiology, Vol. 38, No. 1, July 1979, pp. 1-6.

A method for the concentration of Salmonella from water is described. As is done with enterovirus, Salmonella bacteria were concentrated from water by pH 3.5 adsorption on and pH 9.5 elution from 8-mum porosity microfiber glass filter tubes. This method worked in less than 30 min and S. typhimurium was inactivated only slightly in spite of rapid pH variations (pH 3.5 to 9.5). The retention by the filters stems from two phenomena--a low retention in the microfiber glass labyrinth for small filtered volumes, and a high retention by adsorption at pH 3.5 for any filtered volume (experiments done with 15- and 80-L samples). Addition in tap water of trivalent ions like Al did not increase Salmonella adsorption. In most of the trials, Salmonella recovery varied from 42 to 93 percent. Enterovirus and Salmonella may be concentated from the same water sample by this procedure.

Chappelle, E.W. et al., "Rapid, Quantitative Determination of Bacteria in Water", Patent-4 385 113, 1983, U.S. Patent Office, Washington, D.C.

A bioluminescent assay for ATP in water-borne bacteria is made by adding nitric acid to a water sample with concentrated bacteria to rupture the bacterial cells. The sample is diluted with sterile, deionized water, then mixed with a luciferase-luciferin mixture and the resulting light output of the bioluminescent reaction is measured and correlated with bacteria present. A standard and a blank also are presented so that the light output can be correlated to bacteria in the sample and system noise can be substracted from the readings. A chemiluminescent assay for iron porphyrins in water-borne bacteria is made by adding luminol reagent to a water sample with concentrated bacteria and measuring the resulting light output of the chemiluminescent reaction.

Fleisher, J.M. and McFadden, R.T., "Obtaining Precise Estimates in Coliform Enumeration", Water Research, Vol. 14, No. 5, 1980, pp. 477-483.

The use of proper experimental design and improved techniques of coliform enumeration for improved quantitative use of coliforms as water quality indicators is demonstrated. Water samples from a moderately polluted beach were analyzed by the membrane filtration technique and the density estimates analyzed by a nested analysis of variance. This analysis showed that less than or equal to 95 percent of the variation in coliform density could be ascribed to actual temporal changes in density when replicated determinations are made. By decreasing the variance of the density estimates, replicated determinations allowed the construction of confidence intervals about the estimates which were considerably smaller than with single determinations. The need for increased precision has significance for routine water quality surveys as well as for research situations but is often overlooked. The theoretical superiority of the membrane filtration technique over the multiple tube fermentation method was discussed.

Hirn, J. and Pekkanen, T.J., "The Stability of Simulated Water Samples for the Purpose of Bacteriological Quality Control", Vatten, Vol. 33, No. 3, 1977, pp. 318-323.

When the properties of the improved formate lactose glutamate storage medium and a modified nutrient broth were compared for the preservation of the coliforms and the fecal coliforms at room temperature (20°C) they were almost equal, but the modified nutrient broth was much easier to prepare. The improved formate lactose glutamate storage medium had a better freeze protective effect on the microbes studied when stored at -35°C. A modified sodium azide broth was well suited to the preservation of the fecal streptococci. A testing of the Finnish water control laboratories was done using the modified media.

Jeffers, E.L., "Method and Apparatus for Continuous Measurement of Bacterial Content of Aqueous Samples", PAT-APPL-SN-891 247, 1978, U.S. Patent Office, Washington, D.C.

The methods and apparatus for automatically and continuously making quantitative determinations of the bacteria present in water samples such as waste water, effluent or fresh water are presented. A bacteria adenosine triphosphate was used to determine the number of live bacteria present and the iron porphyrin assay to determine the total number of bacteria alive and dead present in the sample.

Joshi, S.R. et al., "An Indigenous Membrane Filter Medium for Enumeration of Coliform in Water", Indian Journal of Environmental Health, Vol. 20, No. 1, Jan. 1978, pp. 29-35.

The membrane filter (MF) technique for the bacteriological analysis of water is simple, quick and less cumbersome than the multiple tube dilution technique. A medium for the enumeration of coliforms from water by MF technique was developed with indigenously available peptones and was compared with imported M. endobroth medium for its efficacy to detect coliforms from water samples of different pollutional loads. The data was statistically analyzed and is presented. The indigenous medium compared favorably with that of the endobroth and the cost of the medium was also low. There was no significant difference between the mean counts of the two media both at the 95 and 99 percent level of significance, despite differences in pollutional load within the water samples and in different sources.

Lesar, D.J. and Standridge, J.H., "Analytical Note: Sampling and Storage Methodology for the Detection of Sulfate-Reducing Bacteria", Journal of the American Water Works Association, Vol. 71, No. 7, July 1979, p. 406.

A study was undertaken to determine the length of time and the conditions under which a water sample to be tested for sulfate reducers could be stored or shipped before analysis without affecting the results. The use of glass or plastic containers caused no statistical difference in the number of sulfate reducers detected. Bacterial counts decreased with storage time, although some bacteria were detectable for less than or equal to 1 month after sample collection. At low levels of contamination, samples should be run within 48 hr of collection. Storage temperature had no effect on results in the range 2°C to 25°C, nor did incubation temperature variations in the range 35°C to 57°C.

Manja, K.S., Maurya, M.S. and Rao, K.M., "A Simple Field Test for the Detection of Faecal Pollution in Drinking Water", Bulletin of the World Health Organization, Vol. 60, No. 5, 1982, pp. 797-801.

A comprehensive field investigation in several parts of India has revealed that the presence of coliforms in drinking water is associated with hydrogen sulfide-producing organisms. This paper describes a simple, rapid, and inexpensive field test for the screening of drinking water for faecal pollution, based on the detection of hydrogen sulfide. The new test showed good agreement with the standard most probable number (MPN) test. It proved highly successful in the field when it was used to detect faecal pollution and to monitor water quality during an outbreak of waterborne hepatitis A infection in the city of Gwalior.

Martins, M.T., Alves, M.N. and Sanchez, P.S., "Comparison of Methods for Pseudomonas Aeruginosa Recoveries from Water", Environmental Technology Letter, Vol. 31, No. 9, 1982, pp. 405-410.

A comparative study of Pseudomonas aeruginosa enumeration methods was carried out on different types of water. Two membrane filter (MF) and two Most Probable Number (MPN) multiple tube techniques were compared, with typical colonies being biochemically identified for confirmation. Results were analyzed for sensitivity and specificity. MF techniques presented higher specificity than MPN procedures, and for salt water samples the MF procedure based on MPA-B medium presented higher sensitivity than the MF procedure based on MPA medium. Pseudomonas aeruginosa, a ubiquitous organism, has been isolated from surface waters and soil, and is capable of infecting plants, animals and man. Its presence in water is frequently attributed to man as the human enteric tract is the main reservoir of the P. aeruginosa of environmental contamination.

Reasoner, D.J., Blannon, J.C. and Geldreich, E.E., "Rapid Seven-Hour Fecal Coliform Test", Applied and Environmental Microbiology, Vol. 38, No. 2, Aug. 1979, pp. 229-236.

A rapid 7-hr fecal coliform (FC) test for the detection of FC in water has been developed. This membrane filter test utilizes a lightly buffered lactose-based medium (m-7-hr FC medium) combined with a sensitive pH indicator system. FC colonies appeared yellow against a light purple background after incubation at 41.5°C for 7-7.25 hr. The mean verified FC count ratio (6-hr FC count/24-hr FC count) for surface water samples was 1.08. The mean FC count ratio (7-hr FC count/24-hr FC count) for unchlorinated wastewater ranged from 1.95 to 5.05. Verification of yellow FC colonies from m-7-hr FC medium averaged 97 percent. Data from field tests on Lake Michigan bathing beach water samples showed that unverified 7-hr FC counts averaged 96 percent of the 24-hr FC counts. The 7-hr FC test is suitable for the examination of surface waters and unchlorinated sewage, and could serve as an emergency test for detection of sewage or fecal contamination of potable water.

Schillinger, J.E., Evans, T.M. and Stuart, D.G., "Rapid Determination of Bacteriological Water Quality by Using Limulus Lysate", Applied and Environmental Microbiology, Vol. 35, No. 2, Feb. 1978, pp. 376-382.

The Limulus lysate assay, which was used to measure the endotoxin content of water samples from the East Gallatin River (Montana) watershed, accurately predicted the degree of bacterial contamination as measured by coliform, enteric, gram-negative, and heterotrophic bacteria. The firm-clot method was less sensitive and less reproducible in detecting endotoxin than the spectrophotometric modification of the Limulus lysate assay, and bound endotoxin, determined in the same manner, is a better measure of the endotoxin associated with bacterial cells than is total endoxotin. The usefulness of this assay, in addition to sanitary conditions, might be extended to assessing Fe bacteria, sulfate-reducing bacteria, and a number of sheathed bacteria in pipes and water mains. Additional refinement of the method is necessary to make it amenable to routine water quality testing, but a successful development of this rapid test would be a significant advancement in fields such as water microbiology and pharmaceutics.

APPENDIX G

BIOLOGICAL MONITORING

Bingham, C.R. et al., "Grab Samplers for Benthic Macroinvertebrates in the Lower Mississippi River", Misc. Paper E-82-3, July 1982, U.S. Army Engineer Waterways Experiment Station, Vicksburg, Mississippi.

The use of any one single type and size of existing grab sampler for gathering representative benthic macroinvertebrate samples from the various habitats within rivers is impractical, if not impossible. Adjusting one sampling gear (grab sampler) to the existing habitat and program objectives appears to be the best approach. Although macroinvertebrate relative catch efficiency (catch by area inscribed by cocked sampler) varies among sampler types, the catch efficiency of a single type varies with substrate, current regime, and bottoms contours to a greater extent. Data from 24 grab samples, eight each taken with the Standard Ponar, Petite Ponar, and Shipek grabs from a backwater habitat were statistically compared using one-way analysis of variance. Total grabs produced 5696 organisms and 24 distinct taxa. Results showed no difference in catch efficiency among grab samplers for distinct taxa, total densities, and densities of the dominant taxa: Lirceus sp., Ilyodrilus templetoni, Hexagenia sp., Limnodrilus cervix, Limnodrilus hoffmeisteri, Sphaerium sp., and immature Iliodrilus. The Petite Ponar grab captured significantly fewer immature Limnodrilus. This was attributed to clumped distribution of these worms rather than difference in gear type. Analysis of the cumulative percent composition of newly acquired species showed that second and additional replicates of each grab type accounted for 10 percent or less of the total standing crop, as also found in marine substrate types.

A single grab per station is sufficient to characterize the dominant benthic macroinvertebrate standing crop community and is, therefore, recommended for survey level studies. The Shipek grab is the preferred grab in habitats with strong currents; rough bottom morphology; and sand, gravel, or firm clay substrates. Ponar-type grabs are preferred for softer substrates under lower current velocities. This includes most backwater-type habitats, e.g., river borders, abandoned channels, oxbow lakes, and dike fields under low flow conditions. The choice between Petite Ponar and the Standard Ponar grabs depends upon the project design, but should be made with the following facts in mind. A single Standard Ponar grab samples a larger surface area than a single Petite Ponar grab; therefore, it provides a better representation of the immediate (station) benthic macroinvertebrate community than does the Petite Ponar grab. However, almost twice as many Petite Ponar samples can be taken and processed as Standard Ponar samples with similar effort. Greater numbers of stations dispersed over an area gives a better areal representation. Greater number of replicates provide smaller experimental error and, therefore, better support data for statistical inferences.

Cairns, J., Jr., Dickson, K.L. and Westlake, G.F., "Biological Monitoring of Water and Effluent Quality", ASTM Special Technical Publication 607, 1976, American Society for Testing Materials, Philadelphia, Pennsylvania.

Papers are presented on reflections on laboratory testing and ecological guidelines, biotic monitoring, an industrial view of biotic monitoring, evaluation of an automated biological monitoring system at an industrial site, an electronic system to monitor the effects of changes in water quality on fish opercular rhythms, a biological monitoring system employing rheotaxis of fish, rotatory-flow technique for testing fitness of fish, an automatic system for rapid detection of acute high concentrations of toxic substances in surface water using trout, and rapid assessment of water quality using the fingernail clam. Other topics include monitoring and analytical techniques for the study of locomotor responses of fish to environmental variables, a laser-based optical filtering system to analyze samples of diatom communities, automated measurement of river productivity for eutrophication prediction, biological monitoring in the activated sludge treatment process, the importance of monitoring change, research related to biological evaluation of complex wastes, methods for hypothesis testing and analysis with biological monitoring data, and time as a factor in biomonitoring estuarine systems with reference to benthic macrophytes and epibenthic fishes and invertebrates.

Cairns, J., Jr. and Gruber, D., "A Comparison of Methods and Instrumentation of Biological Early Warning Systems", Water Resources Bulletin, Vol. 16, No. 2, Apr. 1980, pp. 261-266.

Rapid biological information systems using aquatic organisms to monitor water and wastewater quality have only recently begun to develop technologically for practical on-site applications. One feasible approach monitors the ventilatory behavior of fish to assess the quality of drinking water supplies and industrial wastewater discharges. A comparison of the basic strategies of the various biological monitoring systems making use of this concept is presented. The applications and potential utilization of these systems are discussed.

Hellawell, J.M., Biological Surveillance of Rivers: A Biological Monitoring Handbook, 1978, Water Research Center, Stevenage, England.

This 10-chapter monograph deals with water quality and pollution surveys, including monitoring objectives, water standards and criteria, field surveys, sampling strategies (for macroinvertebrates, fish, benthic organisms, macrophytes, etc.), data analysis and biotic indices, comparisons of data-handling methods, presentation and interpretation of survey results, and sources of additional help (keys for taxa identification, mathematical tables, bibliographic references, etc.).

Jacobs, F. and Grant, G.C., "Guidelines for Zooplankton Sampling in Quantitative Baseline and Monitoring Programs", EPA/600-3-78/026, Feb. 1978, Virginia Institute of Marine Science, Gloucester Point, Virginia.

Methods of zooplankton sampling and analysis for quantitative baseline and monitoring surveys are described and evaluated. Zooplankton exhibit wide spatial, diurnal, and seasonal variations which, along with gear bias and capture avoidance, complicate data collecting and subsequent assessment of relationships. This study concludes: (1) baseline studies require more frequent sampling and closely spaced stations than do monitoring studies; (2) sampling locations can be further apart in homogeneous waters, while in heterogenous coastal or estuarine water sites should be more closely spaced and sampling conducted more frequently; (3) sampling sites can be selected by means of a grid overlaid on the study area, though transects may be used with study areas which cover great distances and when ship time is limited; (4) in pollution studies a series of transect lines radiating from a single source may be advisable; (5) pumping systems are an expensive but efficient means of capturing microzooplankton; a rate exceeding 150 liters/min is necessary to minimize avoidance; (6) nets with mouth openings of 50-100 cm diameter are recommended for most groups of mesozooplankton; (7) in areas of high plankton density, a 333-micrometer mesh is preferable; (8) specimens are generally best preserved in 4 percent buffered formaldehyde; and (9) the Folsom and Burrell splitters are suggested. Statistical methods of analysis are discussed, and a bibliography is provided.

Kingsbury, R.W. and Rees, C.P., "Rapid Biological Methods for Continuous Water Quality Monitoring", Effluent and Water Treatment Journal, Vol. 18, No. 7, July 1978, pp. 319-331.

Biological "early warning systems" monitor behavioral or physiological responses of single or groups of organisms. The organisms should be able to remain under test without suffering metabolic or behavioral disorders, be reasonably sensitive and able to react rapidly to a wide range of pollutants, have readily measurable and quantifiable reactions, and be unaffected by outside influences other than water quality. Fish are the most popular organisms, but due to adaptation phenomena are best suited to demonstrate rapidly acute aberrations in water quality. They may indicate water quality deterioration through alteration of swimming movements, cough reflex, breathing rate, and heart beat. Algae, bacteria, protozoa, and macroinvertebrates have also been employed. High levels of toxin can be detected fairly rapidly in fish and low levels are detectable within 24 hr using catheters. Of the other biomonitors, only the bacterial nitrifying column offers both rapidity and continuity of monitoring while maintaining ease of preparation. Biomonitors should supplement rather than replace physicochemical methods.

National Technical Information Service, "Fresh Water Mussels and Water Quality--1970--June, 1983 (Citations from the NTIS Data Base)", 1983, U.S. Department of Commerce, Springfield, Virginia.

This bibliography contains citations concerning the use of molluscs to monitor water quality, and the effects of water quality upon the distribution and abundance of molluscs. Molluscs include mussels, clams, and other bivalves. Water quality is affected by sedimentation, eutrophication, heavy metals, pesticides and other toxicants. The studies include both the investigation of effects and the specific reports for identified rivers, lakes and impoundments. Contains 79 citations fully indexed and including a title list.

Roline, R.A. and Miyahara, V.S., "Evaluation of the Algal Assay Bottle Test", REC-ERC-80-1, 1979, U.S. Bureau of Reclamation, Denver, Colorado.

The algal assay bottle test has been used in the study of water quality and nutrient enrichment on many water and power projects. Water samples from 12 sites in the Western United States were used in this evaluation of the test. Actual cell counts and dry mass measurements were the best indicators of algal production. In most studies, the growth of the test alga Selenastrum capricornutum Printz responded positively to an increase in nutrient concentration; however, the selected test alga may not be the best test organisms for all fresh waters. Some inhibition of growth may have been due to high salinity or toxic elements such as heavy metals or herbicides in the water systems.

Slooff, W., "Biological Monitoring Based on Fish Respiration for Continuous Water Quality Control", Second International Symposium on Aquatic Pollutants, 1977, National Institute for Water Supply, Amsterdam, Netherlands.

The detection limits of a monitoring system which records fish respiration patterns using dual external electrodes are delineated for 12 compounds and compared to the corresponding LC50 values for Salmo gairdnerii and Brachydania rerio. The response limit of this sensor ranges between 1 percent (Cd) and 30 percent (acrylonitrile, cyanide, lindane) of the 48-hr LC50. The detection capacity of the monitor system depends mainly on the mode of action of the toxicants. Fast responses of low concentrations will generally be observed for those pollutants which act more or less directly on the respiratory system. Other chemicals, e.g., carcinogenic compounds or other compounds which require bioactivation into reactive metabolites before exerting a toxic action, may pass unnoticed in hazardous quantities. More responses were obtained during the dark interval than during the light interval, possibly because the breathing rates were generally lower and often less variable during the dark interval than during the light interval resulting in close upper and lower critical values at night. The sensitivity of the fish to the toxicants may also be subjected to a circadian periodicity. This system might be helpful to control water quality in effluents and surface waters in addition to chemical-physical monitors.

States, J.B. et al., "A Systems Approach to Ecological Baseline Studies", FWS/OBS-78/21, Mar. 1978, U.S. Fish and Wildlife Service, Fort Collins, Colorado.

This handbook describes a systematic approach to planning and conducting a "holistic" study of selected ecosystem components and functions, which are significant with regard to energy development projects in the Western United States. Techniques of ecological systems analysis are described, and the manual explains how to build a conceptual ecosystem model and use it to plan a baseline study. A glossary of key terms and an annotated bibliography are included.

Stout, G.E. et al., "Baseline Data Requirements for Assessing Environmental Impact", IIEQ-78-05, May 1978, Institute for Environmental Studies, University of Illinois, Urbana-Champaign, Illinois.

This study has developed a guide that may be used by technical personnel to perform an integrated baseline evaluation of changes in the total environment--in plants, soils, and animals (including man)--that is needed for a factual pinpoint assessment. The methodology outlined in this guide requires substantial resources both in manpower and funds. The management and evaluation of the survey instrument should be performed by a qualified organization. Comprehensive ecosystem evaluation requires an interdisciplinary team of scientists. As a result, the execution of a baseline impact assessment requires considerable planning, funding, and evaluation.

Ward, D.V., Biological Environmental Impact Studies: Theory and Methods, 1978, Academic Press, New York, New York.

This book shows how the time resources presently devoted to descriptive empirical efforts and guessing of impacts can be redirected to focused manipulative experimentation so that the predictability of biological changes is greatly improved. The author specifies how this type of biological environmental impact can be approached and accomplished. The book makes three major contributions: (1) to present the idea and some examples of manipulative rather than descriptive ecological studies to managers and government agents concerned with requesting and reviewing environmental impact assessment studies; (2) to aid the biologist performing the impact study to review rapidly a battery of approaches and methods that are applicable to manipulative impact assessment studies; (3) as an aid to training biology students, many of whom will be employed to perform environmental impact studies. Chapter headings include Environmental Impact Analysis, The Field Survey: Preliminary System Analysis, Modelling the System, The Field Experiment, Laboratory Studies, Some Examples, and Conclusions.

Weber, C.I., "Biological Field and Laboratory Methods for Measuring the Quality of Surface Waters and Effluents", EPA-670/4-73-001, 1973, U.S. Environmental Protection Agency, Cincinnati, Ohio.

This manual was developed within the National Environmental Research Center - Cincinnati to provide pollution biologists with the most recent methods for measuring the effects of environmental contaminants on freshwater and marine organisms in field and laboratory studies which are carried out to establish water quality criteria for the recognized beneficial uses of water and to monitor surface water quality.

Weiss, C.M., "Evaluation of the Algal Assay Procedure", EPA/600/3-76-064, 1976, U.S. Environmental Protection Agency, Corvallis, Oregon.

Evaluation of the algal assay bottle test and its relationship to the trophic state or nutrient levels of surface waters was examined in 44 lakes impoundments, and rivers in North Carolina in 345 separate assay sets. Of particular concern was the evaluation of the significance of the pretreatment procedure, autoclaving or filtration, upon growth of the reseeded alga in relationship to the original water quality. A limnological data profile was developed for each of the bodies of water sampled. A data processing procedure was used to establish the relationship between water quality data and algal cell density, chlorophyll a and productivity.

APPENDIX H

LOCATION OF WATER QUALITY MONITORING STATIONS
(ABSTRACTED FROM SANDERS, 1980)

INTRODUCTION

The location of a permanent sampling station is probably the most critical design factor in a monitoring network which collects water quality data. If the samples collected are not representative of the water mass, the frequency of sampling as well as the mode of data interpretation and presentation becomes inconsequential. Nevertheless, criteria to establish station locations for representative sampling has received relatively little attention from agencies responsible for the collection and dissemination of water quality data.

There are three levels of design criteria in the discussion of sampling station location which are considered. They are: the macrolocation--river reaches which will be sampled within the river basin; the micro-location--station location relative to outfalls or other unique features within a river reach; and representative locations--points in the river's cross section from which grab samples will provide a lateral profile of the stream. The macrolocation is essentially a river reach assigned a sampling station which draws samples from the river's transect, while the designation of a microlocation remains site specific, i.e., a transect would be selected following examination of specific characteristics of the macrolocation. The macrolocation is specified systematically so elements of personal bias are largely removed from the selection. The utility of data collected from a water quality network will be largely dependent upon the consideration given to each level in the planning of the data collection network. The macrolocation will be a function of the specific objectives of the sampling agency while the microlocation, which defines a zone of complete mixing, is a function of the hydraulic and mixing characteristics of the stream.

SAMPLING STATION LOCATIONS

Location of sampling stations is closely tied to the number of station locations available which, in turn, is often a function of the number of samples to be taken at each station. Thus, prior to a detailed evaluation of sampling station location, the general sampling frequency ranges and the total number of stations available needs to be determined. These should not be too difficult to estimate if the purpose of the network is well defined, and the budget and sampling costs are known.

If the water quality monitoring network covers a political jurisdiction (e.g., a state), some allocation of monitoring resources must be made between the different basins being managed. If the network has two or more objectives, allocations must be made between the objectives.

Allocation by Objective

Allocation between objectives can be achieved by determining the emphasis the decision makers place on the different objectives. If an agency has two monitoring objectives (e.g., determine water quality trends and detect streams in violation of their standards), then the policymakers can be questioned (e.g., by a questionnaire) to determine their opinion as to how much of the monitoring resources should be allocated to each objective.

Once the objectives are determined, the monitoring resources (samples and, thus, stations) can be divided. Returning to the above example to illustrate, stations allocated to detect trends can generally be assigned fewer samples than stations allocated to detect streams in violation of their standards.

The major point of the above discussion is that monitoring objectives must be defined if a rational basis for allocating stations (resources) to an objective is to be made. If the network has one objective, the problem is simplified, but this is rarely the case.

Allocation by Basin

As alluded to above, different areas of a river basin may be classified as more important than others based on potential water quality management problems. Such a classification may be a basis for allocating monitoring resources. A basin that has little population and development would probably not need an extensive amount of monitoring. On the other hand, a basin with a large population and high industrial activity would probably need a more extensive water quality management program and, therefore, more monitoring than the undeveloped basin.

Differences in basins can be quantified via an index approach. A management need index could include population, industrial activities, agricultural use, and other factors. The monitoring resources, in parallel with management resources, could then be allocated among the basins using the index as the basis.

IDENTIFYING WATER QUALITY SAMPLING SITES IN A RIVER BASIN--MACROLOCATION

If the intent of sampling were to study a limited portion of a river, as required to discover the immediate downstream effects of a given discharge (synoptic surveys), the placement of sampling transects might not be too critical to the generation of information descriptive of the reach in question. However, when the intent of sampling is to monitor whole rivers or river basins, any sample taking must be preceded by a thoughtful selection of reaches in the river basin to be sampled.

Basic approaches to identifying macrolocations include one based upon percentage areal coverage and the second based upon the density of some indicator of population, where the density of population is considered to correspond to the likelihood of polluting episodes and overall discharge of pollutants. In the percentage areal coverage approach stations are placed systematically to generate data on water quality in the entire river basin. This approach lends itself to characterizing trends in a river basin.

Another approach is to designate sampling sites according to some logical basis: for example, to concentrate sampling near known sources of pollution. While this approach may come closest to generating data which reflect the quality of a stream as it varies with longitudinal position, it will nevertheless reflect bias unless a rational systematic procedure for designating sampling stations is employed.

Selection of Reaches and Tributaries in Which to Sample

This procedure for specifying sampling station locations systematically subdivides the river network into portions which are relatively equal in terms of the number of contributing tributaries. The method was proposed for use in defining a water quality sampling network having a primary objective to detect, isolate and identify a source of pollution.

Each exterior tributary or link contributing to the mainstem of a river is assigned a magnitude of one. The overall number of exterior tributaries included in the procedure is a function of the scale of map used. An exterior tributary is considered to be a stream fed by no other defined tributaries or having a specified minimum mean discharge. A stream which is formed by the intersection of two exterior tributaries becomes a second order tributary. Continuing downstream in the same manner, a section of river formed by the intersection of two upstream tributaries would have a magnitude equal to the sum of magnitudes of the intersecting streams. At the mouth of the system the magnitude of the final river section will be equal to the number of contributing exterior tributaries. The centroid of the basin is defined by dividing the magnitude of the final stretch of the river by two. The first centroid divides the river network into two approximately equal portions and a new centroid may be found for each. The location designated by the first division is referred to as a first-hierarchy sampling reach. Division of the entire network into quarters defines second-hierarchy sampling reaches. Successive subdivisions define increasing levels of hierarchy. The transects within these reaches at which samples are to be taken are termed sampling stations.

MICROLOCATION

Once the macrolocations of sampling stations have been specified according to the objectives of the monitoring network, further definition of the location is required to insure that water quality samples are truly representative of the section of river monitored.

The representativeness of a water quality sample is a function of the uniformity of the sample concentrations in a river's cross sectional areas. Wherever the concentration of a water quality variable is independent of depth and lateral location in a river's cross section, the river at that point is completely mixed and is defined as a microlocation.

Well mixed zones (microlocation) in a river for representative water quality sampling can be defined, given that several assumptions will apply. By assuming that a pollutant distribution from an instantaneous point source is Gaussian in both the lateral and vertical transect and applying classical image theory, a theoretical distance from an outfall to a well mixed zone in a straight, uniform river channel can be determined. This mixing distance is a function of: 1) mean stream velocity; 2) location of the point source; and 3) the mean lateral and vertical turbulent diffusion coefficients.

There are several models available being functions of the mixing coefficients which have been shown to apply for predicting a zone of relatively complete mixing. One is an expression for a mixing distance utilizing the solution to the steady-state, two-dimensional, advection and dispersion equation. Assuming that most streams are shallow enough so that complete vertical mixing is assured in a relatively short distance, the following relationship from the two-dimensional solution can be used to predict the mixing distance to a point where concentration variation in the cross section does not exceed ten percent.

$$L \geq 0.075 \ \frac{w^2 u}{D_y}$$

where:

L = mixing distance

w = width of channel

u = mean stream velocity

D_y = lateral turbulent diffusion coefficient

In a more general relationship, the mixing distance to a point of uniform concentration in the cross section could be calculated by:

$$L = \frac{1}{2\alpha_1^2} \frac{w^2 u}{D_y}$$

where:

α_1 = constant associated with the location of a point source and the degree of uniformity of the concentration gradient.

The following expression was derived for a mixing distance based on the channel geometry and the point of tracer injection relative to the channel midpoint. This was done primarily to find the point in the river where flow measurement by the dilution method is applicable (uniform concentration in the cross section).

$$L = \frac{K_1}{0.02} \frac{w^2}{d}$$

where:

K_1 = a function of the lateral turbulent diffusion coefficient, the tracer injection point and the width of the straight river

d = depth of flow.

When estimating flow by the dye dilution method, the United States Geological Survey recommended the use of the following equations to locate sampling points downstream from dye injection points at the center of the stream and near the bank, respectively.

$$L = 1.3 \ u \ \frac{w^2}{d}$$

$$L = 2.6 \ u \ \frac{w^2}{d}$$

It should be noted that the mixing distances are a function of the square of the river width, the lateral turbulent diffusion and a constant.

Assuming that the distribution of a pollutant initiated from a point source is theoretically a Gaussian distribution, Sanders defined mixing distances in a uniform, straight river channel for both complete lateral mixing and vertical mixing with the following equations:

$$L_y = \frac{\sigma_y^2 u}{2D_y}$$

$$L_z = \frac{\sigma_z^2 u}{2D_z}$$

where:

L_y = mixing distance for complete lateral mixing,

L_z = mixing distance for complete vertical mixing,

σ_y = distance from farthest lateral boundary to point in section,

σ_z = distance from farthest vertical boundary to point of injection.

Because the theoretical mixing distance for complete vertical mixing of the tracer in most streams will be less than the theoretical mixing distance associated with lateral mixing, the mixing distance for providing a uniform pollutant distribution over the entire stream cross section will be equal to the larger of the two mixing distances, that is, the mixing distance associated with lateral mixing.

Example: Will a sampling point 1.3 miles downstream from an outfall located at midstream and middepth of a river having a width of 200 ft and a depth of 10 ft be completely mixed so that only one sampling point in the lateral transect is required? It is assumed there is negligible thermal and density stratification of the effluent and that the mean stream velocity is 3 ft/sec and the lateral and vertical turbulent diffusion coefficients are 0.5 ft²/sec and 0.01 ft²/sec, respectively.

$$L_y = \frac{\sigma_y^2 u}{2D_y} = \frac{(100 \text{ ft})^2 (3 \text{ ft/sec})}{2(0.5 \text{ ft}^2/\text{sec})}$$

$$L_y = 30,000 \text{ ft} = 5.68 \text{ miles}$$

$$L_z = \frac{(5 \text{ ft})^2 (3 \text{ ft/sec})}{2(0.01 \text{ ft}^2/\text{sec})} = 3750 \text{ ft}$$

$$L_z = 0.71 \text{ miles}$$

It would appear that the river at that location is completely mixed vertically but not laterally.

SELECTED REFERENCE

Sanders, T.G., editor, "Principles of Network Design for Water Quality Monitoring", July 1980, 312 pp., Colorado State University, Ft. Collins, Colorado.

APPENDIX I

AUTOMATIC SAMPLING AND REMOTE SENSING

Briggs, R., Page, H.R.S. and Schofield, J.W., "Improvements in Sensor and System Technology", International Workshop on Instrumentation and Control for Water and Wastewater Treatment and Transport Systems, 1977, Water Research Centre, Stevenage Lab, London, England.

Water quality monitoring networks should be based on a flexible modular system of telemetric and data processing modules capable of containing appropriate mathematical models and deploying only the minimum of robust and reliable sensors necessary to achieve a given objective. The following significant determinants are considered: temperature, DO, organic matter, suspended solids, Cl-, F-, nitrate, ammonia, heavy metals, trace organics, and toxicity. Sensors should be accurate, reliable, and require little maintenance. They should be interchangeable with others of the same type without recalibration, and unaffected by changes in other variables. They should be inexpensive, and provide an output easily connected to the telemetry system. Use of duplicate sensors in a mode such that one measures sample and the other standard, coupled with the ability to switch streams to confirm or deny unusual quality data, provide automatic calibration, and extend duty cycles, has enabled satisfactory performance to be achieved from many existing sensors. Computers can be used to calculate functions of several determinants or pollution indexes so that alarms can be based on a number of factors taken together. The physicochemical monitoring scheme in the River Wear uses dual sensors for the measurement of DO, temperature, turbidity, organic matter, and ammonia. Operation of the monitors is controlled by a minicomputer at the treatment plant. The dual sensor system was used to monitor DO in settled sewage at Washington Wastewater Treatment Works. During design, construction, and commissioning, it is advisable to spend effort on staff training and motivation.

Brinkhoff, H.C., "Continuous On-Stream Monitoring of Water Quality", International Workshop on Instrumentation and Control for Water and Wastewater Treatment and Transport Systems, 1977, Philips Science and Industry Division, London, England.

Continuously operating on-site monitors were used to measure pH, Cl-, ammonium, DO, redox potential, temperature, conductivity, and TOD at two locations in the Netherlands and one in the Federal Republic of Germany. Only Cl- measurement accuracy was worse than the average laboratory analysis. On the basis of qualitative and quantitative information provided by the on-site measurements, it is possible to identify a phenomenon approximately using only the conventional parameters. In a further identification, automatic sampling can play an important role. The sampling should be initiated whenever a present threshold value is exceeded, to allow for detailed laboratory analyses. Continuous monitoring should be carried out at the following points: influents to drinking water supplies; effluents from sewage treatment plants; effluents from major industries; confluences of streams; and border passages. Mobile units would give the control system additional flexibility. Possible data presentation formats include registration of the frequency and duration that preset limit values are exceeded; determination of mean daily, weekly, or monthly values with SDs; and composition of histograms.

Cavagnaro, D.M., "Automatic Acquisition of Water Quality Data. Volume 1. 1970-1975 (A Bibliography with Abstracts)", NTIS/PS-76/0670, 1977, National Technical Information Service, U.S. Department of Commerce, Springfield, Virginia.

   The abstracts included in this report cite the techniques used to obtain continuous water quality data as well as general system management and planning studies. Site selection, monitor design, performance evaluation, government needs, and new techniques, including the use of NASA's LANDSAT, are covered. This bibliography contains 154 abstracts.

Cavagnaro, D.M., "Automatic Acquisition of Water Quality Data. Volume 2. 1976-July, 1978 (A Bibliography with Abstracts)", NTIS/PS-78/0887/6, 1978, National Technical Information Service, U.S. Department of Commerce, Springfield, Virginia.

   This bibliography cites techniques used to obtain continuous water quality data. General system management and planning studies are also included. The citations cover site selection, monitor design, performance evaluation, Government needs, and new techniques including the use of NASA's LANDSAT. This updated bibliography contains 216 abstracts.

Cavagnaro, D.M., "Automatic Acquisition of Water Quality Data. Volume 2. 1976-1977 (A Bibliography with Abstracts)", NTIS/PS-79/1054/0, 1979, National Technical Information Service, U.S. Department of Commerce, Springfield, Virginia.

   Studies on techniques used to obtain continuous water quality data are cited. The citations cover site selection, monitor design, performance evaluation, government needs, and new techniques including the use of LANDSAT satellites. They also include general system management and planning studies. This updated bibliography contains 190 abstracts, none of which are new entries to the previous edition.

Cavagnaro, D.M., "Automatic Acquisition of Water Quality Data. Volume 3. 1978-August, 1979 (A Bibliography with Abstracts)", NTIS/PS-79/1055/7, 1979, National Technical Information Service, U.S. Department of Commerce, Springfield, Virginia.

   This bibliography cites techniques used to obtain continuous water quality data. General system management and planning studies are also included. The citations cover site selection, monitor design, performance evaluation, Government needs, and new techniques including the use of LANDSAT satellites. This updated bibliography contains 138 abstracts, 112 of which are new entries to the previous edition.

Cavagnaro, D.M. and Hundemann, A.S., "Automatic Acquisition of Water Quality Data. 1978-August, 1980 (Citations from the NTIS Data Base)", PB80-815772, 1980, National Technical Information Service, U.S. Department of Commerce, Springfield, Virginia.

   This bibliography cites techniques used to obtain continuous water quality data, with emphasis on biological and satellite remote sensing techniques. Monitor design and performance and general systems management and planning studies are included. This updated bibliography contains 168 citations, 30 of which are new entries to the previous edition.

Davis, P.E., "A Study of the Reliability of Continuous Water Quality Monitoring", PB-228 872/8, 1973, National Technical Information Service, U.S. Department of Commerce, Springfield, Virginia.

An investigation was made of the reliability of continuous water quality monitoring on Fort Loudoun Reservoir. Two continuous monitoring systems were used, and the parameters of temperature, dissolved oxygen, pH, and conductivity were considered. Manual cross sectioning was used to determine the various values in the stream. At both stations, the data taken from the monitor was found to provide an accurate prediction of the conditions of the stream. A comparison of samples taken at the pumps and at the monitors showed that no appreciable change in any of the parameters was induced by the intake systems.

Eagleson, K.W., Morgan, E.L. and McCollough, N., "Water Quality Monitoring Using an Automated Biomonitor and NASA's GOES Satellite", Journal of Tennesse Academy of Science, Vol. 53, No. 2, Apr. 1978, p. 76.

A biological monitor was developed using fish breathing rates as a measure of biological response to physical-chemical water quality changes. These changes alter the expected breathing rates and are used in interpreting complex water quality fluctuations. A remote capacity is necessary if an effective regional monitoring network is to be established. This requirement was met by using NASA's GOES satellite for data transmission. Simultaneous physical-chemical data is provided by the National Park Service using similar satellite data transmission. Data is then statistically tested for correlations between physical-chemical fluctuations and biological responses. The system has potential for wilderness monitoring, identification of regional nonpoint source run-off pollutants, and watershed management.

Fisher, P.D. and Siebert, J.E., "Integrated Automatic Water Sample Collection System", Journal of the Environmental Engineering Division, American Society of Civil Engineers, Vol. 103, No. EE4, Aug. 1977, pp. 725-728.

Twelve integrated automatic water sampling systems were assembled and installed at stream gaging stations. The systems consist of water level recorder, level-threshold detector, cam-activated microswitch, 12 V DC power supply, an auxiliary electronic controller, and a water sample collector. As the stream height rises, the level recorder pulley and cam rotate; the microswitch cam-follower drops off the cam edge, closing the normally closed switch contacts. This action switches on the power to the sample collector and auxiliary electronic controller. Sample collection is then dictated by the electronic controller. The sample collector shuts off automatically after collection of the last sample. The systems have 3 distinct advantages over systems in which gage height data and water samples are collected independently. Water quality and stream flow are no longer treated independently, and the total discharge of a particular pollutant in a given time interval may be determined more precisely. Needless water samples are not collected since samples are collected only in response to a hydrologic event. Since the auxiliary electronic controller can be programmed to collect water samples at one rate on the rising or falling hydrograph, a more precise determination of stream water quality is possible without increasing the number of water samples collected.

Grana, D.C. and Haynes, D.P., "Remote Water Monitoring System", PATENT-4 089 209, 1977, U.S. Patent Office, Washington, D.C.

A remote water monitoring system is described that integrates the functions of sampling, sample preservation, sample analysis, data transmission and remote operation. The system employs a floating buoy carrying an antenna connected by lines to one or more sampling units containing several sample chambers. Receipt of a command signal actuates a solenoid to open an intake valve outward from the sampling unit and communicates the water sample to an identifiable sample chamber. Such response to each signal receipt is repeated until all sample chambers are filled in a sample unit. Each sample taken is analyzed by an electrochemical sensor for a specific property and the data obtained is transmitted to a remote sending and receiving station. Thereafter, the samples remain isolated in the sample chambers until the sampling unit is recovered and the samples removed for further laboratory analysis.

Koehler, F.A., "Simple Sampler Activation and Recording System", Journal of the Environmental Engineering Division, American Society of Civil Engineers, Vol. 104, No. EES, Oct. 1978.

A device that resulted in automated sample procurement and simultaneous hydrograph time-stage notation was developed as part of a project to assess the impact of rural nonpoint sources on an areawide basis. Five systems were assembled and installed at stream monitoring stations. Significant improvements from the use of this system as opposed to time interval sampling are as follows: samples are taken in response to hydraulic changes in the stream, when greatest changes in the stream chemistry are expected; repetitive samples under steady-state flow conditions are eliminated, reducing the analytical laboratory workload; streamflow and water chemistry are measured concurrently, allowing more precise calculation of constituent transport; no prior knowledge or assumptions of the hydraulic response time is necessary to adequately characterize a runoff event; and all streamflow regimes are sampled, runoff events as well as baseflow recession, in proportion to the expected changes in water chemistry.

Lillesand, T.M., Scarpace, F.L. and Clapp, J.L., "Photographic Quantification of Water Quality in Mixing Zones", NASA-CR-137268, 1973, University of Wisconsin, Madison, Wisconsin.

A method was developed to quantitatively delineate waste concentrations throughout waste effluent mixing zones on the basis of densitometric measurements extracted from aerial photography. A mixing zone is the extent of a receiving water body utilized to dilute a waste discharge to a concentration characteristic of a totally mixed condition. Simultaneously acquired color infrared photography and suspended solids water samples were used to quantitatively delineate the mixing zone resulting from the discharge of a paper mill effluent. Digital scanning microdensitometer data was used to estimate and delineate suspended solids concentrations on the basis of a semi-empirical model. Photographic photometry, when predicated on a limited amount of ground sampling, can measure and delineate mixing zone waste distributions in more detail than conventional surface measuring techniques. The method has direct application to: (1) the establishment of definite and rational water quality guidelines; (2) the development of sampling and surveillance programs for use by governmental and private agencies; and (3) the development of design and location criteria for industrial and municipal waste effluent outfalls.

National Technical Information Service, "Automatic Acquisition of Water Quality Data. 1978-January, 1982 (Citations from the NTIS Data Base)", PB82-804790, 1982, U.S. Department of Commerce, Springfield, Virginia.

This bibliography cites techniques used to obtain continuous water quality data, with emphasis on biological and satellite remote sensing techniques. Monitor design and performance and general systems management and planning studies are included. This bibliography contains 297 citations, 129 of which are new entries to the previous edition.

Rogers, R.H., "Application of Landsat to the Surveillance and Control of Lake Eutrophication in the Great Lakes Basin", NASA-CR-143409, 1975, National Aeronautics and Space Administration, Washington, D.C.

By use of distilled water samples in the laboratory, and very clear lakes in the field, a technique was developed where the atmosphere and surface noise effects on LANDSAT signals from water bodies can be removed. The residual signal is dependent only on the material in water, and can be used as a basis for computer categorization of lakes by type and concentration of suspended material. Several hundred lakes in the Madison and Spooner, Wisconsin area were categorized by computer techniques for tannin or nontannin waters and for the degree of algae, silt, weeds, and bottom effects present. When the lakes are categorized as having living algae or weeds, their concentration is related to the enrichment or eutrophication of the lake.

Thiruvengadachari, N.G. et al., "Some Ground Truth Considerations in Inland Water Surveys", International Journal of Remote Sensing, Vol. 4, No. 3, 1983, pp. 537-544.

Ground data requirements are more stringent and difficult in aerial water quality surveys than in land surveys, as a result of the dynamic nature of water quality as well as the complex energy-water interaction given rise to noise components in the signal measured by the airborne sensor. This paper presents some of the considerations influencing the ground truth strategy adopted in recent NRSA (National Remote Sensing Agency) water quality studies. Various aspects such as depth of sampling, time lags, location of sampling sites on remotely sensed data, elimination of spectral noise components and planning of redundancy in ground truth are addressed. Careful collection, preservation, storage and analysis of water samples can contribute significantly to reliable water quality mapping from aerial altitudes.

Wallace, J.W., Lovelady, R.W. and Ferguson, R.L., "Design, Development, and Field Demonstration of a Remotely Deployable Water Quality Monitoring System", EPA-600/4-81-061, 1981, U.S. Environmental Protection Agency, Hampton, Virginia.

A prototype water quality monitoring system is described which offers almost continuous in situ monitoring.   The two-man portable system features: (1) a microprocessor controlled central processing unit which allows preprogrammed sampling schedules and reprogramming in situ; (2) a subsurface unit for multiple depth capability and security from vandalism; (3) an acoustic data link for communications between the subsurface unit and the surface control unit; (4) eight water quality parameter sensors; (5) a nonvolatile magnetic bubble memory which prevents data loss in the event of power interruption; (6) a rechargeable power supply sufficient for 2 weeks of unattended operation; (7) a water sampler which can collect samples for laboratory analysis; (8) data output in direct engineering units on printed tape or through a computer compatible link; (9) internal electronic calibration eliminating external sensor adjustment; and (10) acoustic location and recovery systems.

Welby, C.W., "Use of Multispectral Photography in Water Resources Planning and Management in North Carolina", UNC-WRRJ-76-115, 1976, University of North Carolina, Chapel Hill, North Carolina.

A series of multispectral aerial photographic missions were flown over the Chowan River of North Carolina to study and evaluate possible uses of multispectral photography in planning for and management of North Carolina's water resources.   Flights were also made over parts of the Cape Fear River and Northeast Cape Fear. This report describes how the technique might be used to trace pollutants associated with turbid water and how water draining from the swamps might be traced in larger water bodies.   The technique has potential usefulness for monitoring phytoplankton buildups and can be used even if only limited groundtruth information is available. Areas of apparently differing water quality were monitored by the multispectral photography both in the Chowan River and in the Northeast Cape Fear River. A technique was developed by which the difference between the extinction depths of a red quadrant and a white quadrant on a modified Secchi disc can be used with water sampling and analysis to obtain biomass concentrations indirectly. This relationship can be one of the groundtruth measurements used to establish relationships between the reflectance characteristics of a water body and various water quality parameters directly or indirectly.

APPENDIX J

WATER QUALITY AND BIOLOGICAL INDICES

Ball, R.O. and Church, R.L., "Water Quality Indexing and Scoring", Journal of the Environmental Engineering Division, American Society of Civil Engineers, Vol. 106, No. EE4, Aug. 1980, pp. 757-771.

The desirability of developing water quality indices, which has been frequently cited, was examined. The development of currently available indices was reviewed, and a distinction was made between the indexing and scoring of water quality samples. It was pointed out that many indexes are, in fact, scoring methods, and that the mathematical basis for the scoring methods are not always consistent with the objectives of scoring. A general mathematical approach to scoring was described, and the necessity of considering the uniformity as well as the average quality of the water was described. The development and use of "true" indices was briefly reviewed, and recommendations for future indexing approaches were presented.

Booth, W.E., Carubia, P.C. and Lutz, F.C., "A Methodology for Comparative Evaluation of Water Quality Indices", 1976, Worcester Polytechnic Institute, Worcester, Massachusetts.

The relative usefulness of water quality assessment indices is examined by use of a methodology that compares and evaluates their performance characteristics. The method so devised demonstrates the evaluation of two indices by identifying their assumptions and limitations. Evaluation of the National Sanitation Foundation (NSaF) index shows that it is only suitable to assess the overall water quality as characterized by nine physicochemical parameters. It is somewhat insensitive to pollution problems exhibited in only one of the parameter's values. Since the parameters quality ratings, and weights that are used are fixed, they cannot vary to account for the ultimate use of the index, data availability, new discoveries, or conditions associated with different geographic locations. Applications of the NSaF index are extremely limited when attempting to identify or analyze specific pollution problems as they may affect a water use. Conversely, the Harkins index is best used when assessing trends in a specific pollution problem area. The index has no set parameter requirements and is therefore very flexible in application since only those parameters of interest to the user need be included in the computations.

Dunnette, D.A., "A Geographically Variable Water Quality Index Used in Oregon", Water Pollution Control Federation Journal, Vol. 51, No. 1, Jan. 1979, pp. 53-61.

An Oregon Water Quality Index (OWQI), which takes into account water quality differences resulting from geographical characteristics of separate basins was developed to provide a simple, concise, and valid method for expressing the significance of regularly generated laboratory data. The trend-monitoring value of the index was demonstrated for two Willamette River stations. Correlations among this and other proposed indexes averaged 0.87. Yearly and seasonal variations in water quality were quantitized and averaged 88.9 and 78.9 OWQI units for the higher and lower water quality stations respectively over the period 1971-76. Calculated rates of change in water quality were +0.68 and +0.91 OWQI units/yr for the two stations for 1971-76. The OWQI is now used routinely in Oregon's primary station sampling program to recognize water quality trends.

House, M. and Ellis, J.B., "Water Quality Indices: An Additional Management Tool", Water Science and Technology, Vol. 13, No. 7, 1981, pp. 413-423.

A comparison of the geometric weighted formulation of the Scottish Development Department (SDD) water quality index with the National Water Council (NWC) and Thames Water Authority (TWA) classifications showed 76 percent and 81 percent agreement, respectively. This suggests that, with modifications, the SDD index could be successfully used to monitor changes in water quality. In its present form it is most accurate in good quality waters and less accurate in low quality waters. There are several advantages to adopting a water quality index: (1) it reduces a large amount of data to a single index; (2) it reduces subjective judgments; (3) it provides more information on water quality; (4) it allows greater detail in describing specific river reaches and in showing small changes in quality; (5) it reports specific quality rather than qualitative approximations; and (6) it is more easily understood by laymen.

Inhaber, H., Environmental Indices, John Wiley and Sons, Inc., 1976, New York, New York.

This book provides a general summary of environmental indices, beginning with a discussion of concepts and economic indices. Chapters are included on air quality indices, water indices, land indices, biological indices, aesthetic indices, and other environmental indices.

Ott, W.R., Environmental Indices--Theory and Practice, 1978, Ann Arbor Science Publishers, Inc., Ann Arbor, Michigan, 1978.

This book systematically describes environmental index systems along with principles for their design, application and structure. Chapter 1 introduces the reader to environmental data, presenting simple communicative approaches such as environmental quality profiles. It also describes national monitoring activities. Chapter 2 presents a new conceptual framework that is designed to embrace nearly all existing environmental indices, allowing the behavior of different index structures to be compared and probed in detail. Chapter 3 concentrates on air pollution indices, using the conceptual framework introduced in Chapter 2 to analyze and compare published air pollution indices. Chapter 3 also gives a detailed summary of the historical evolution and scientific basis for the Pollutant Standards Index (PSI), which has been developed for uniform application throughout the United States. Computational aids (equations, tables and nomograms) for applying PSI to actual air quality data are included. Chapter 4 covers water pollution indices, using the theoretical framework and concepts from Chapter 2 to examine currently used water pollution indices; it also presents design principles for an ideal water quality index and discusses a candidate index structure. In both Chapters 3 and 4, the current air and water index usage patterns in the United States are described in detail. Finally, Chapter 5 presents conceptual approaches, such as Quality of Life and environmental damage functions, that extend beyond the traditional fields of air and water pollution.

Polivannaya, M.F. and Sergeyeva, O.A., "Zooplankters as Bioindicators of Water Quality", Hydrobiological Journal, Vol. 14, No. 3, 1978, pp. 39-43.

The purpose of this study was to conduct a comparative evaluation of methods of biological analysis of water quality by using, as an example, zooplankton of the Dnieper River and the Kanev reservoir (USSR). The material consisted of samples collected in the river in 1969 and 1972, and in the reservoir in 1975. The results indicate that pollution leads to less varied species composition of the zooplankton. The first to suffer are crustaceans and among them, the nonpredator forms. Polyphagous insects (Asplanchnidae, Brachionus calyciflorus, Brachionus angularis and predator Cyclopoidae) tend to maintain themselves in the plankton. The data indicate an abrupt dip of the number of zooplankton dominants in river zones subjected to pollution. Age composition, replenishment and loss, fertility, morphometric characters of populations of leading species under normal conditions and during disturbances associated with pollution, may serve as supplementary indices of water quality that are calculated in terms of zooplankton.

Yu, J.K. and Fogel, M.M., "The Development of a Combined Water Quality Index", Water Resources Bulletin, Vol. 14, No. 5, Oct. 1978, pp. 1239-1250.

Use-oriented benefits and treatment costs analysis has been incorporated into a water quality index to show an economically optimized concentration for the treatment of the pollutants and the resulting water quality. This combined water quality index can be used in decision making at the Federal and local government levels. Five major parameters, i.e., coliforms, nitrogen, phosphorus, suspended solids, and detergents, have been considered for the municipal wastewater. With each higher level of improvement, the treatment costs increase accordingly and the benefits associated with the reuse of this treated wastewater will increase also, but not for the nutrient removal in agricultural use. The optimal concentration is determined when the marginal costs equal the marginal benefits. The combined water quality index is the combination of the maximum net benefits and the water quality index of the optimized residual concentrations. This water quality index is zero dollars for the Tucson region in this study. The possible reclaimed use of municipal wastewater is for agricultural irrigation and recreational lakes for the Tucson region.

INDEX